AGRICULTURAL CHEMICALS
BOOK I

Insecticides, Acaricides and Ovicides

BY
W. T. THOMSON

1989 Revision

THOMSON PUBLICATIONS
P.O. BOX 9335
FRESNO, CA 93791

Library of Congress Catalog No. 64-24795
Printed in the United States Of America
ISBN - Number 0-913702-33-1

THIS BOOK FOR REFERENCE ONLY

READ THE LABEL CAREFULLY

TABLE OF CONTENTS

Page 1

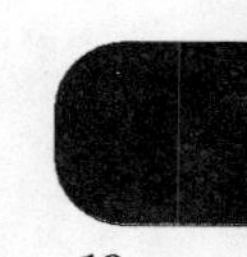

Page 19

Page 61

Page 117

Page 155

Page 175

Page 221

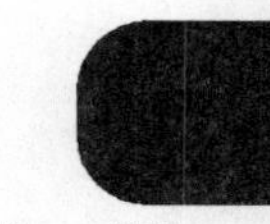

Page 253

INTRODUCTION

This book is designed as a helpful guide to all the available agricultural chemicals and their uses. Farmers, agronomists, pesticide salesmen, pest control operators, greenskeepers, conservationists, county agents, research workers, and many others should find this a handy reference for frequent use.

In this book the author has attempted to list and describe the most widely used agricultural chemicals marketed in the world today. Many of the compounds listed are not available commercially, as they are still in the experimental stages; however, by mentioning them, the life and usefulness of this book will be prolonged.

Since research is continuing on most of the chemicals available in the agricultural field, a check with your farm advisor, county agent, pesticide supplier, agricultural college, or other leading authority would be advisable before using the chemicals as recommended. The facts have been presented as they appear to the author, but this information is continually changing as knowledge is gained on the use of these materials.

HOW TO USE THIS MANUAL

The following indicate what is to be found under each heading for the individual chemicals listed. The chemicals are not listed alphabetically, but placed in groups of related compounds. All names, both chemical and common, are listed in the indexes appearing in the front of the book.

NAMES: An Attempt has been made to list all the usual names by which a chemical is known. Since this manual is designed for use mainly by the non-technical individual, the common and trade names are used primarily. The most popular trade or common name is listed first, while the remaining are placed at random, not according to usage. The official common name, if the chemical has one, is italicized.

Immediately under the names appears the structural formula, if such a formula exists, followed by a written chemical formulas, for the sake of simplicity usually only one appears.

ORIGIN: The company which has done most of the development work on the compound is listed, followed by the year the compound was patented or put on the market. The chemical company mentioned may or may not have patented the compound, but it has done much of the basic research connected with developing it. On compounds developed outside the United States, the company which is licensed to market the chemical in this country is mentioned. On some of the older chemicals, only the principle basic producers are listed.

Since many companies may formulate the same chemical compound, readers should not be led to the believe that the company name is the only producer. The author is by no means trying to endorse certain companies' compounds.

TOXICITY: The LD_{50} values of the different compounds are listed under this heading. Since they vary considerably with different tests conducted by different personnel, the lowest value found in the literature is given. The values are listed as the acute oral LD_{50} of the technical material, usually recorded in milligrams per kilogram of body weight (mg/kg). As most toxicology work is done with white albino rates, the value listed is that of the toxicity on them unless otherwise stated. Other information on toxicity may be mentioned.

FORMULATIONS: This category explains the forms in which a chemical is marketed. They are usually wettable powders (WP), emulsifiable concentrates (EC), oil solutions, granules, dusts, or aerosols. Since these formulations are constantly changing, the reader is advised to check into the different formulations available from his supplier in his particular locality.

PHYTOTOXICITY: Many pesticides, when applied to some plants, show detrimental side effects. The author has endeavored to list as many plants injured by a certain compound as it was possible to find. This changes in many cases due to the advent of new, less phytotoxic formulations. Weather conditions in a certain locality may also be a factor. Many of the chemicals are still in the experimental stage, so phytotoxicity data on them may not be complete.

USES: Since this manual was designed primarily for use in the United States, the plants and animals for use on which the different pesticides are registered by the Environmental Protection Agency and the United States Department of Agriculture are listed under this heading. This can serve as a guide for usage outside of the United States. An attempt was made to list all the registered uses at the time of publication, but many uses may have been deleted or added since then. Your farm advisor, county agent, or chemical supplier can also give you proper verification before applying to a questionable usage. A good rule to remember is: don't use a pesticide if the desired usage is not referred to on the chemical label unless the feasibility of such usage is confirmed by a reliable source. *Read the label carefully.*

Experimental uses, so designated, are mentioned for some of the newer compounds. The compounds may or may not be registered for such uses within the next few years.

RATES: The extremes of high and low dosages are stated in most cases on a per-acre basis and on a per-100-gallons-of-water basis. The average dosages used should normally fall in between these rates. Your supplier or county agent can give you much more accurate information on this for your specific locality or situation.

IMPORTANT DISEASES CONTROLLED OR PREVENTED: The heading is self-explanatory in that only the pests considered by the author to be of major importance are mentioned. In the interest of simplicity and space, not all the pests a certain chemical will control or prevent could be listed.

APPLICATION: This, at its best, is only a general guide to the application of these compounds. As locality, weather, rates, crops, etc., vary greatly, only a general recommendation can be given. Authorities in the reader's locality can give much more specific recommendations.

PRECAUTIONS: This is self-explanatory in that the possible hazards connected with the use or application of the specific compound are mentioned.

RELATED MIXTURES: Mixtures containing the compound previously mentioned are listed here, together with the company which produces them. Due to space restrictions, not all such mixtures could be mentioned, so only a few with registered trade names were selected. They are mentioned merely to let the reader know they are available and to avoid confusion of the different names.

RELATED COMPOUNDS: Pesticides which are closely related to the above compounds are listed. These, so listed, are used to a very limited extent, and do not warrant a full page. Only a brief description is given.

STATEMENT OF WARRANTY

NOTICE

Since new pesticides are constantly being introduced, a revision of this manual will be available when the number of such chemicals warrants it. As of this writing, there are a series of four manuals available: Agricultural Chemicals Book I, Insecticides and Acaricides; *Agricultural Chemicals Book II - Herbicides; Agricultural Chemicals Book III - Miscellaneous Chemicals, Fumigant, Growth Regulators, Repellents, and Rodenticides; and Agricultural Chemicals Book IV - Fungicides.* More information may be obtained by writing the publisher.

TRADEMARKS

The material in this book has been assembled from a multitude of labels, bulletins, instruction sheets, etc., published by the various companies for the public's information in the use of their products. For the sake of simplicity, reference to and use of registered trademarks has been eliminated. This should, by no means, indicate the absence of proprietary right on the use of such words. Also, by the omission of certain trade names, either unintentionally or from lack of space, the author should not be considered to be endorsing only the companies whose brand names are listed.

NAMES, COMPOUNDS & MIXTURES

A

B

C

D

E

F

G

H

M

N

O

P

Q

R

S

T

U

V

W

Y

Z

CYCLO COMPOUNDS

NAMES

ALDRIN, ALDREC, ALDRON, ALGRAN, HHDN, SOILGRIN

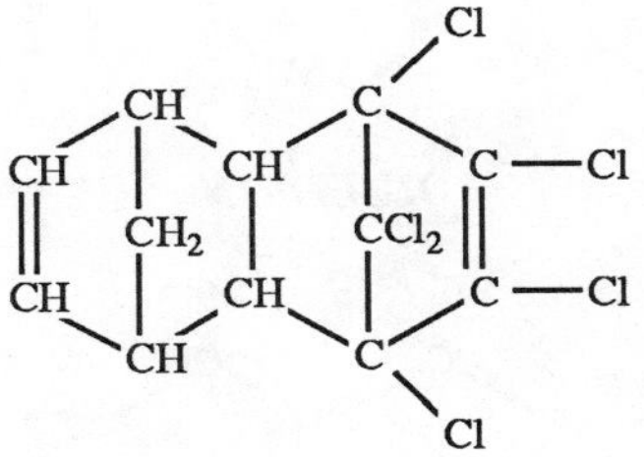

Hexachlorohexahydro-endo, exo-dimethanonaphthalene

TYPE: Aldrin is a chlorinated-hydrocarbon insecticide with contact and stomach-poison action.

ORIGIN: J. Hyman and Co. Licensed to be developed by Shell Chemical Company, 1948. Produced by Shell Int'l. outside the U.S.

TOXICITY: LD_{50}-38 mg/kg. Absorbed through the skin.

FORMULATIONS: 4 EC, granules, WP, and solutions.

PHYTOTOXICITY: Non-phytotoxic to crops when used at the recommended rates. The solvent may be phytotoxic to some plants in certain formulations.

USES: No longer sold in the U.S. Used outside the U.S. on corn, potatoes, small grains, sorghum, sugar beets, tobacco, sugarcane, bananas, orchards, and other crops, and for termite control.

IMPORTANT PESTS CONTROLLED: Ants, cutworms, armyworms, crickets, Diabrotica, wireworms, fleahoppers, grasshoppers, Japanese beetles, chinch bugs, leaf miners, slugs, snails, sowbugs, spittlebugs, thrips, and many others.

RATES: Applied 1/2-5 lb actual/A.

APPLICATION: Used as a soil insecticide, seed treatment, and on the plant foliage.

PRECAUTIONS: Not for sale or use in the U.S. Do not use on animals. Do not contaminate lakes, streams, or ponds. Do not let livestock graze on treated crops. Do not use in greenhouses. Persists in the soil. Toxic to fish.

ADDITIONAL INFORMATION: Long residual effects. Nonsystemic. Compatible with other insecticides. Valuable as a soil insecticide for termites. No off-flavor in crops has been reported. Considered harmless to soil microorganisms.

NAME

DIELDRIN

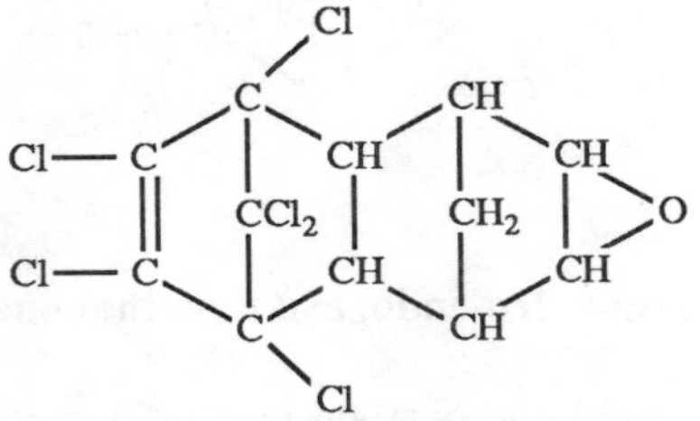

Cl
C CH
Cl— C CH CH
CCl_2 CH_2 O
Cl— C CH CH
C CH
Cl

Hexachloroepoxyoctahydro-endo, exo-dimethanonaphthalene

TYPE: Dieldrin is a chlorinated-hydrocarbon insecticide with contact and stomach-poison activity.

ORIGIN: J. Hyman and Co. Licensed to be developed by Shell Chemical Co., 1948. Produced by Shell Int'l. outside the U.S.

TOXICITY: LD_{50}-40 mg/kg. Absorbed through the skin.

FORMULATIONS: EC, WP, dusts, and granules.

PHYTOTOXICITY: Non-phytotoxic to crops when used at the recommended rates. Some crops may be sensitive to the solvents of certain formulations.

USES: No longer used in the U.S. Used on a number of crops outside the U.S.

IMPORTANT PESTS CONTROLLED: Ants, cutworms, armyworms, loopers, chiggers, chinch bugs, fleahoppers, crickets, Diabrotica, Drosophila, earwigs, wireworms, grasshoppers, flies, Japanese beetles, leaf miners, lygus, mosquitoes, wasps, roaches, slugs, snails, sowbugs, webworms, spittlebugs, termites, thrips, ticks, and many others.

RATES: Applied at 1/2-5 lb actual/A.

APPLICATION: Used outside the U.S. as a soil insecticide, seed treatment, and on the plant foliage. Used for termite control and as a public health insecticide.

PRECAUTIONS: Not for sale or use in the U.S. Do not apply directly to animals, or graze livestock on treated crops. Do not contaminate lakes, streams, or ponds. Toxic to bees. Do not use treated seed for feed, food, or oil purposes. Toxic to fish.

ADDITIONAL INFORMATION: Long residual effectiveness. A nonsystemic insecticide. Compatible with many other insecticides and fungicides. Persists in the soil. Shows no harmful effects to soil microorganisms, nor does off-flavor appear in crops growing in treated soils. Widely used in Africa in the desert locust control program.

NAME

ENDRIN

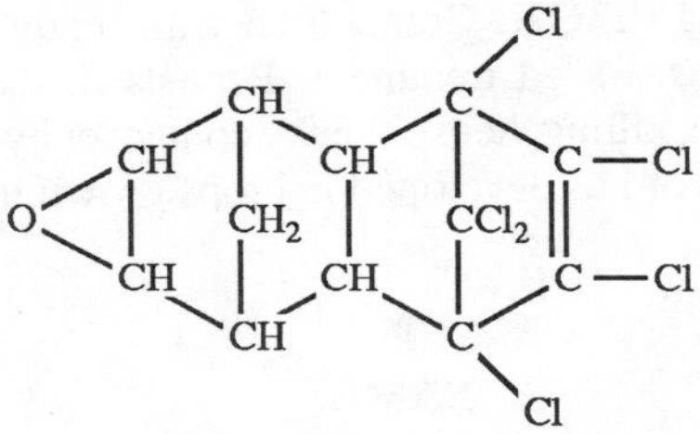

Cl
CH C
CH CH C—Cl
O CH$_2$ CCl$_2$
CH CH C—Cl
CH C
Cl

Hexachloroepoxyoctahydro-endo, endo-dimethanonaphthalene

TYPE: Endrin is a chlorinated-hydrocarbon insecticide with contact and stomach-poison activity.

ORIGIN: J. Hyman and Co. Licensed to be manufactured by Shell Chemical Company and Velsicol Chemical Co., 1950. Produced by Shell Int'l. outside the U.S.

TOXICITY: LD_{50}-7 mg/kg. Absorbed through the skin. Very toxic to fish.

FORMULATIONS: Granules, EC, WP, and baits.

PHYTOTOXICITY: Injury has been reported on corn and cucumbers. Possible phytotoxic effects from certain solvents and additives.

USES: Barley, cotton, oats, ornamentals, rye, sugarcane, and wheat. Also registered for mouse control on orchard floors. Used outside the U.S. on many crops, particularly cotton, rice, and sugarcane.

IMPORTANT PESTS CONTROLLED: Ants, aphids, armyworms, boll weevils, bollworms, loopers, chinch bugs, cornborers, corn earworms, crickets, cutworms, grasshoppers, hornworms, leafhoppers, leaf miners, lygus bugs, spittlebugs, thrips, and many others.

RATES: Applied at 1/4-3/4 lb actual/100 gal of water of 1/8-1 lb actual/A.

APPLICATION: Begin when insects first appear and repeat as often as necessary. Agitate while spraying. Only freshly mixed spray solutions should be used, since separation may occur in the spray tank upon standing. Also used as a soil insecticide.

PRECAUTIONS: No longer producted in the U.S. Workers entering treated fields within 5 days of application should be protected. Do not feed treated crop residue or graze treated fields. Do not contaminate streams, lakes, or ponds. Do not apply directly to animals. Extremely toxic to fish and other wildlife.

ADDITIONAL INFORMATION: Compatible with many other insecticides and fungicides. Used as a seed treatment. Persists in the soil. Used also as a rodenticide on mice, killing them either by contact or by their feeding on treated foliage. No off-flavor has developed in crops grown in treated soils.

NAMES

CHLORDANE, **KYPCHLOR, OCTACHLOR, SYDANE, SYNKLO**

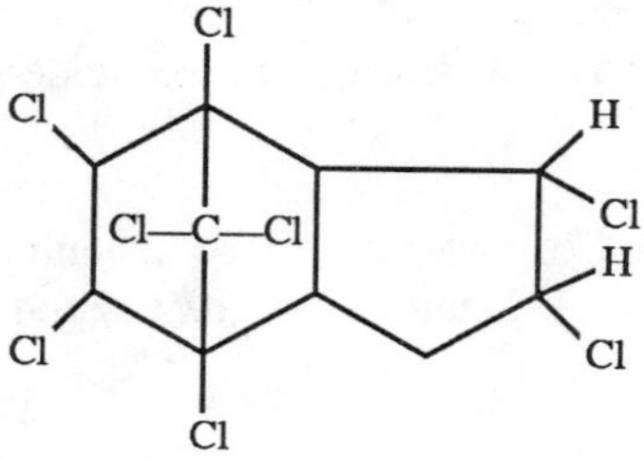

Octachloro-4, 7-methanotetrahydroindane

TYPE: Chlordane is a chlorinated-hydrocarbon insecticide showing contact, stomach-poison, and fumigant action with long residual effects.

ORIGIN: Velsicol Chemical Company, 1945.

TOXICITY: LD_{50}-250 mg/kg. Moderately irritating to the skin. Absorbed through the skin.

FORMULATIONS: 1,2,4,5 and 8 EC, 40 and 50% WP.

PHYTOTOXICITY: Considered Non-phytotoxic when used at the recommended rate. Off-flavor has resulted in potatoes, corn, and plums. High concentrations are injurious to some vegetables. Residues in the soil may depress germination.

USES: No longer sold in the U.S. Used outside the U.S. on numerous crops.

IMPORTANT PESTS CONTROLLED: Ants, cutworms, armyworms, leaf miners, weevils, ticks, loopers, lice, chiggers, chinch bugs, cockroaches, corn earworms, boll weevils, crickets, earwigs, wireworms, flies, fleas, grasshoppers, Japanese beetles, wasps, lygus bugs, mosquitoes, thrips, sowbugs, spiders, spittlebugs, termites, and many others.

RATES: Applied at 1-10 lb actual/A.

APPLICATION: Used as a soil insecticide, seed treatment, and on the plant foliage.

PRECAUTIONS: Do not use in the U.S. Do not feed treated forage or crop residues to livestock. Persists in the soil. Harmful to fish. Incompatible with alkaline solvents and carriers.

ADDITIONAL INFORMATION: Gives residual termite control in the soil for at least 5 years. Compatible with other pesticides. Nonsystemic. Usage in the U.S. is being gradually phased out and eventually will be limited to termite control only.

NAME

HEPTACHLOR

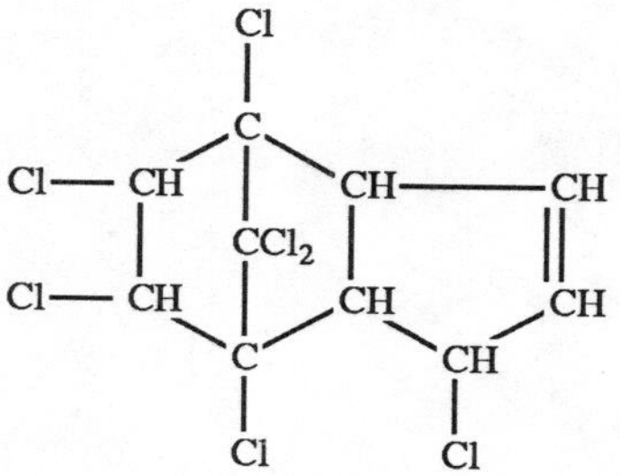

```
               Cl
               |
               C
             /   \
Cl— CH       |     CH———————CH
    |      CCl2    |         ||
    |        |     |         ||
Cl— CH       |     CH        CH
      \      |    /   \     /
         C             CH
         |             |
         Cl            Cl
```

Heptachlorotetrahydro-4, 7-methanoindene

TYPE: Heptachlor is a chlorinated-hydrocarbon insecticide with contact and stomach-poison action, especially effective on soil insects.

ORIGIN: Velsicol Chemical Company, 1948. Produced by other manufacturers outside the U.S.

TOXICITY: LD_{50}-40 mg/kg. Absorbed through the skin.

FORMULATIONS: 2 EC, 3 EC, 25% WP, 2-1/2, 5, 20, and 25% granules, and 1.5-2.5% dusts.

PHYTOTOXICITY: No injury occurs to crops when used at the recommended rates.

USES: No longer sold in the U.S. Used outside the U.S. on numerous crops.

IMPORTANT PESTS CONTROLLED: Ants, spittlebugs, weevils, wireworms, cornborers, rootworms, boll weevils, leafhoppers, cutworms, thrips, Japanese beetles, grasshoppers, plum curculio, and many others.

RATES: Applied at 1/4-4 lb actual/A. For seed treatment use 1/2-1.5 oz actual/100 lb of seed.

APPLICATION: Used as a soil treatment, seed treatment, and on the plant foliage.

PRECAUTIONS: Do not use in the U.S. at this time. Do not mix treated seed with bare hands. Do not feed treated forage to animals. Persists in the soil. Incompatible with alkaline compounds.

ADDITIONAL INFORMATION: Compatible with other pesticides. Used in the transplant water of some crops to control soil insects. May be used in the soil for termite control. Agitate while spraying. Expresses fumigant action.

NAMES

***ENDOSULFAN*, BENZOEPIN, BEOSIT, CHLORTIEPIN, CYCLODAN, MALIX, MELOPHEN, THIFOR, THIMUL, THIODAN, THIONEX, THIOSULFAN, TIONEL, TIOVEL**

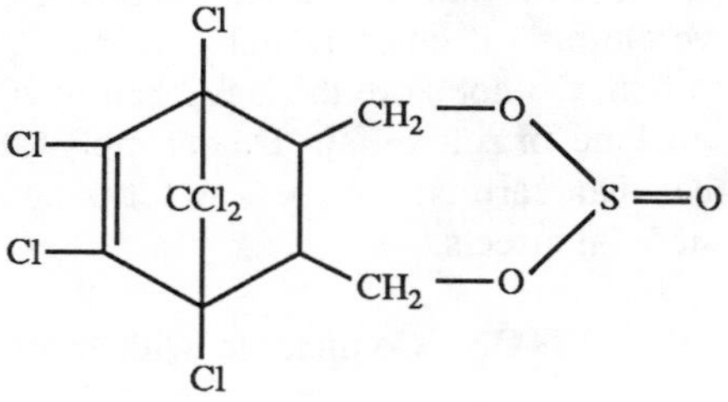

6,7,8,9,10,10-Hexachloro-1,5,5a,6,9,9a-hexahydro-6, 9-methano-2,4,3-benzodioxathiepin-3-oxide

TYPE: Thiodan is a nonsystemic contact and stomach-poison insecticide acaricide.

ORIGIN: Developed in Germany by Hoechst AG, and licensed to be sold in the U.S. by FMC Corp., 1956. Also formulated by other companies.

TOXICITY: LD_{50}-40 mg/kg.

FORMULATIONS: 2 EC, 3 EC, 35 and 50% WP, dusts, and granules.

PHYTOTOXICITY: Not recommended for use on Concord grapes. Some injury has resulted on lima beans and alfalfa. Birch trees are injured. Injury has been noted on geraniums and some varieties of chrysanthemums under greenhouse conditions.

USES: Alfalfa, almonds, apples, apricots, artichokes, barley, beans, blueberries, broccoli, Brussels Sprouts, cabbage, carrots, cauliflower, celery, cherries, citrus, collards, corn, cotton, cucumbers, eggplants, filberts, grapes, kale, lettuce, macadamia nuts, melons, mustard greens, nectarines, oats, peaches, pears, peas, pecans, peppers, pineapples, plums, prunes, potatoes, pumpkins, rice, rye, safflower, southern peas, spinach, squash, strawberries, sugar beets, sugarcane, sunflower, sweet potatoes, tea, tobacco, tomatoes, walnuts, watercress, wheat, and ornamentals. Widely used outside the U.S.

IMPORTANT PESTS CONTROLLED: Aphids, beetles, bollworms, psyllids, tsetse fly, leafhoppers, fleabeetles, stemborers, stinkbugs, boll weevils, loopers, corn earworms, peach twig borers, armyworms, cyclamen mites, and many others.

RATES: Applied at 1/4-1/2 actual/100 gal of water or .2-4 lb actual/A.

APPLICATION: Uniform coverage with common application equipment is necessary. Apply when insects first appear, and repeat as necessary.

PRECAUTIONS: Corrosive to iron. Do not feed treated crop residue to livestock. Wear protective clothing if entering fields within 24 hours of application. Highly toxic to fish. Do not store the 2 EC below 20°F. Incompatible with calcium arsenate, lime, or zinc sulfate. Do not apply to crops to be followed by a root crop other than carrots, potatoes, sweet potatoes, or sugar beets and selective to beneficial insects.

ADDITIONAL INFORMATION: Compatible with most pesticides. Relatively Non-toxic to bees.

NAMES

OXYTHIOQUINOX, CHINOMETHIONAT,
CHINOMETHIONATE, MORESTAN,
QUINOMETHIONATE

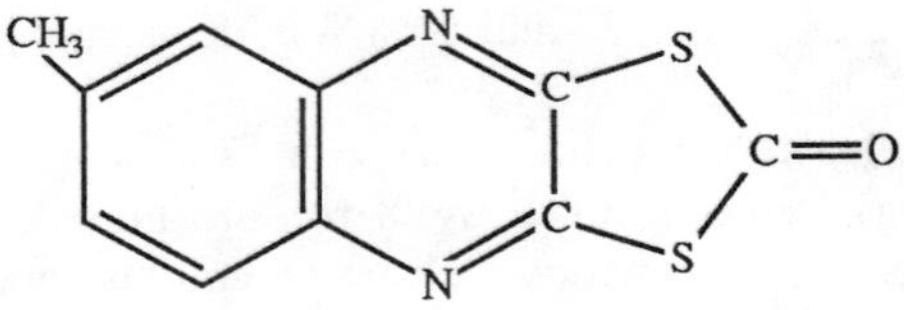

6-Methyl-1,3-dithiolo (4,5-b) quinoxaline-2-one

TYPE: Morestan is an organic-hydrocarbon insecticide-acaricide-fungicide controlling by contact activity.

ORIGIN: Bayer A. G. in Germany, 1962. Sold in the U.S. by Mobay Chemical Corp.

TOXICITY: LD_{50}-1520 mg/kg. May cause skin irritation, but is not a skin sensitizer.

FORMULATIONS: 25% WP.

PHYTOTOXICITY: In the U.S., injury has been reported on Delicious and Winesap apples, D'Anjou pears, some varieties of roses, alders, and other ornamentals.

USES: Apples, apricots, cherries, citrus, nectarines, peaches, pears, plums, prunes, strawberries, walnuts, and ornamentals. Widely used outside the U.S. on these plus vegetables, currents, and gooseberries.

IMPORTANT PESTS CONTROLLED: Mites, pear psylla, and whiteflies. Powdery mildew is also controlled.

RATES: Applied at 1-6 oz actual/100 gal of water or 1/8-1/2 lb actual/A.

APPLICATION: Apply when disease and/or mites appear, and repeat at specified intervals.

PRECAUTIONS: Do not apply with a spreader-sticker. Do not graze treated areas. Do not use in conjunction with summer oils. Toxic to fish. More phytotoxic to roses under high temperature conditions. Do not apply oils after the delayed dormant stage when this material is to be used in the pink stage.

ADDITIONAL INFORMATION: Thorough coverage is necessary for maximum results. Relatively Non-toxic to bees. Active against all stages of mites and against pear psylla. Protective and eradicative activity against powdery mildew.

NAMES

DIENOCHLOR, **MYTEN, PENTAC**

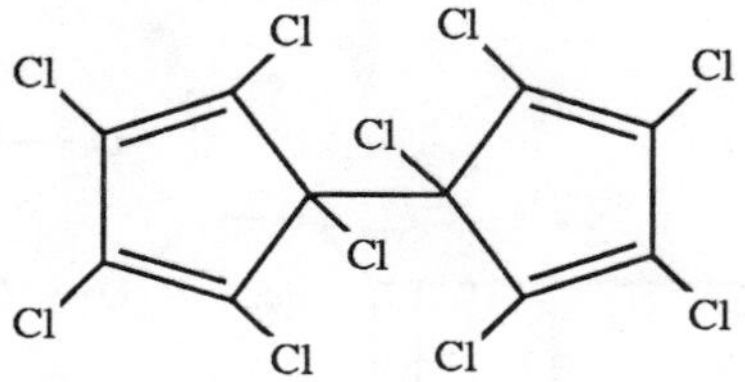

Cl Cl Cl Cl Cl Cl Cl Cl Cl Cl

bis (Pentachlors-2,4-Cyclopentadiene-1-yl)

TYPE: Pentac is a chlorinated-hydrocarbon acaricide expressing long residual activity.

ORIGIN: Hooker Chemical Company, 1960. Now being marketed by Sandoz Crop Protection.

TOXICITY: LD_{50}-3160 mg/kg.

FORMULATIONS: 50% WP. 50% flowable.

PHYTOTOXICITY: Non-phytotoxic when used at the recommended rates.

USES: Greenhouse and outdoor ornamentals.

IMPORTANT PESTS CONTROLLED: Mites.

RATES: Applied at 1/2-1 lb actual/100 gal of water.

APPLICATION: Apply uniformly with common application equipment. Apply before mites attain destructive numbers. Repeat as necessary. The best results are obtained when used as a preventative.

PRECAUTIONS: This compound should not be used on any feed or food crops. Toxic to fish. No insecticidal activity. Use the diluted spray mixture within 24 hours.

ADDITIONAL INFORMATION: A slow-acting material initially, so it should be used in a preventive program or in combination with a knockdown agent. Long residual action. Compatible with most commonly used pesticides. Does not kill beneficial insects. Short lived when used outdoors. Treated mites will stop feeding within hours and die in 1-3-days.

NAMES

HEPTENOPHOS, HOSTAQUICK, RAGADAN

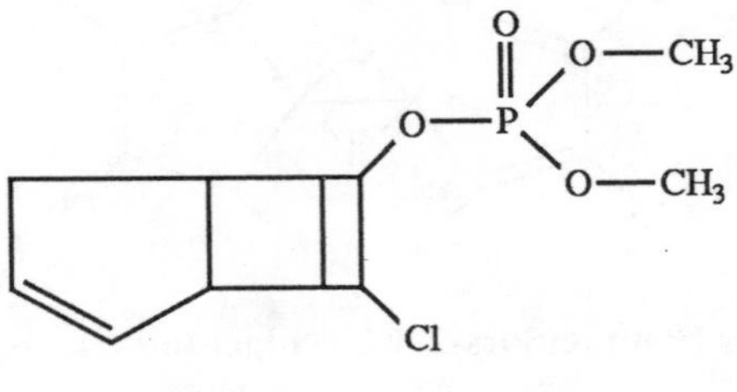

(7-Chloro-bicyclo-(3.2.0)-hepta-2,6-dien-6yl)-dimethyl phosphate

TYPE: Hostaquick is an organic phosphate used as a systemic and contact insecticide.

ORIGIN: Hoechst Ag of Germany, 1970.

TOXICITY: LD_{50}-96 mg/kg.

FORMULATION: 50% EC.

PHYTOTOXICITY: Non-phytotoxic when used as directed.

USES: Outside the U.S on a number of fruit, vegetable, ornamental, and field crops, and on pets and livestock.

IMPORTANT PESTS CONTROLLED: Aphids, fleas, lice, ticks, and Diptera species.

RATES: Applied at .05-.1% ai/ha.

APPLICATION: Apply when insects appear and repeat as necessary.

PRECAUTIONS: Not for sale or use in the U.S.

ADDITIONAL INFORMATION: Quick knockdown with a very short residual. It penetrates the plant tissue and is quickly translocated in all directions.

NAMES

***LINDANE*, AGRONEXIT, GAMAPHEX, GAMMA BHC, GAMMALIN, GAMMEX, ISOTOX, LINDAFOR, LINTOX, NEXIT, NOVIGAM, SILVANOL**

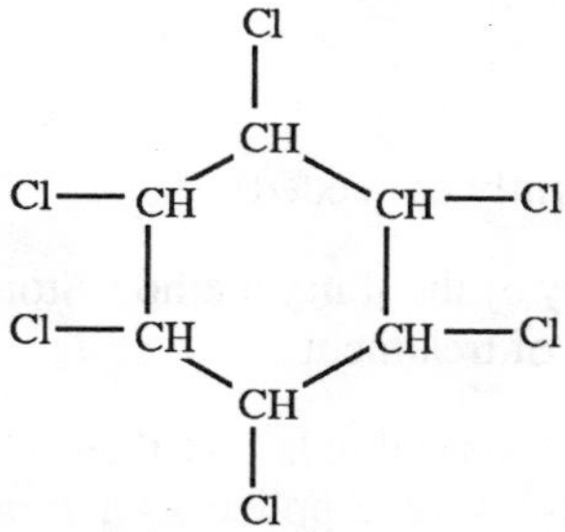

```
              Cl
              |
              CH
           /      \
Cl— CH              CH —Cl
     |               |
Cl— CH              CH —Cl
           \      /
              CH
              |
              Cl
```

Gamma-1,2,3,4,5,6-Hexachlorocyclohexane-at least 99% of the gamma isomer of BHC (Benzene Hexachloride)

TYPE: Lindane is a chlorinate-hydrocarbon insecticide which is an effective stomach, fumigant, and contact poison with relatively long residual effects.

ORIGIN: Chevron Chemical Company, 1945, and ICI of England. Also produced by Rhone Poulenc, Hooker Chemical Co., and Celamerck.

TOXICITY: LD_{50}-88 mg/kg. May be absorbed through the skin.

FORMULATIONS: 25 & 95% WP, 1.7 EC, 100% crystals, 1-3% dusts, 3% smear (screwworm), aerosols. Also sold mixed with other pesticides.

PHYTOTOXICITY: Injury has been reported on potatoes and walnuts. Off-flavor has resulted in some crops. Plant damage will occur if used in excessive amounts.

USES: Alfalfa, apples, apricots, asparagus, avocadoes, barley, beans, broccoli, Brussels sprouts, cabbage, carrots, cauliflower, celery, cherries, clover, collards, corn, cotton, cowpeas, cucumbers, eggplant, flax, grapes, guavas, kale, kohlrabi, lettuce, mangoes, milo, mint, melons, mushroom houses, mustard, nectarines, oats, okra, onions, peaches, peanuts, pears, peas, peppers, pineapples, plums, prunes, pumpkins, quince, radish, rye, safflower, sorghum, soybeans, spinach, squash, strawberries, sudan grass, sugar beets, sunflower, Swiss chard, tobacco, tomatoes, watermelons, wheat, greenhouses, cattle, hogs, goats, sheep, and agricultural premises.

IMPORTANT PESTS CONTROLLED: Aphids, lygus bugs, grasshoppers, spittlebugs, thrips, plum curculio, fleabeetles, ants, leaf miners, cockroaches, armyworms, wireworms, Diabrotica, roaches, flies, mosquitoes, boll weevils, mange mites, termites, and many others.

RATES: Applied at 1/8-lb actual/100 gal of water or 1/4-4 lb actual/A.

APPLICATION:

1. Foliage-Apply uniformly as needed.
2. Seed treatment-Apply by the slurry method. Store treated seed below 70°F and use within 3 months of treatment.
3. Livestock-Do not use on animals less than 3 months old or animals to be slaughtered within 60 days. Applied as a spray not containing over .06% Lindane. Do not use on dairy cattle within 10 days of freshening.
4. Soil-Apply either in a band over the seed row, or broadcast over the entire area.

PRECAUTIONS: More toxic to younger animals. Produces a musty flavor and odor in some food crops, but not as noticeable as BHC. Incompatible with lime sulfur, lime, and calcium arsenate. Corrosive to aluminum. It may persist in the soil for 1 year or more. Toxic to fish, honey bees, and wildlife. Do not use on poultry or poultry houses. Usage in the U.S. is gradually being phased out.

ADDITIONAL INFORMATION: Used to a limited extent in household insecticides, usually combined with pyrethrins. It is 5-10 times as effective on most insects as DDT. Mammals rapidly excrete this compound. The insecticidal properties of BHC are due to one component, the gamma isomer, which in Lindane.

RELATED COMPOUNDS: BHC — An old compound used to a limited extent in the world today. No longer used in the U.S.

NAMES

***PROPARGITE,* COMITE, OMITE, BPPS, RETADOR**

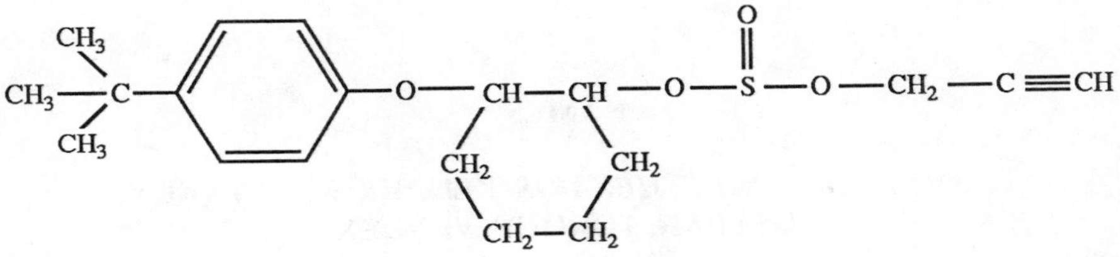

2-(p-tert-Butylphenoxy) cyclohexyl 2-propynyl sulfite

TYPE: Omite is an organic compound being used as an acaricide showing contact and residual activity.

ORIGIN: Uniroyal Chemical Co., 1965.

TOXICITY: LD_{50}-2200 mg/kg. Causes eye damage.

FORMULATIONS: 30% WP, 4% dusts, 6 EC.

PHYTOTOXICITY: Phytotoxic to cotton under 10 inches in height. With the exception of pears and citrus, no injury or fruit blemishes have occurred to date when used as directed. On beans spotting of the pods have occurred.

USES: Apples, almonds, apricots, cherries, corn, grapes, citrus, figs, hops, mint, cranberries, cotton, potatoes, nectarines, strawberries, peaches, pears, plums, prunes, walnuts, peanuts, beans, and sorghum.

IMPORTANT PESTS CONTROLLED: Mites.

RATES: Applies at .375-.75 lb actual/100 gal of water.

APPLICATION: Apply before mites build up to destructive numbers. Apply evenly and thoroughly, repeating as necessary. May be applied by air.

PRECAUTIONS: Do not mix with alkaline materials, oil sprays, or with pesticides containing a large amount of petroleum solvents. Toxic to fish. Do not apply at above 95°F. Do not mix with Diazinon or Imidan for use on nectarines. On citrus leaf injury will occur if applied with oil or if applied 2 weeks before or after an oil application. Do not use in a spray solution with a ph above 10.

ADDITIONAL INFORMATION: Most effective against the motile stages of mites. Nonsystemic. Moderately long residual activity. Effective against phosphate and chlorinated-hydrocarbon-resistant mite strains. No insecticidal activity. Safe to honey bees and allows good survival of predaceous mites. Results are best when the daily temperatures average above 70°F.

NAMES

***FENBUTATIN-OXIDE, HEXAKIS,* BENDEX, NEOSTANOX, OSADAN, TORQUE, VENDEX**

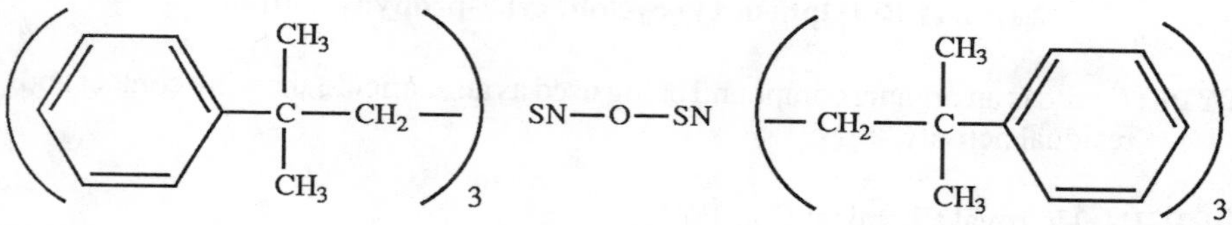

CH_3 CH_3

$(\text{C}{-}CH_2{-})_3$ SN—O—SN $({-}CH_2{-}\text{C}{-})_3$

CH_3 CH_3

Hexakis-(2-methyl-2 phenylpropyl)- distannoxane

TYPE: Vendex is an organic-tin compound used as a contact acaricide.

ORIGIN: DuPont Co., 1974.

TOXICITY: LD_{50}-2630 mg/kg. Irritating to the skin and eyes.

FORMULATION: 50% WP.

PHYTOTOXICITY: Do not use on tangerines, tangelos, and certain grapefruit varieties. Apply only to foliage of chrysanthemums or poinsettias.

USES: Citrus, cherries, cucumbers eggplant, peaches, plums, prunes, strawberries, ornamentals, pecans, walnuts, almonds, grapes, papayas, apples, and pears.

IMPORTANT PESTS CONTROLLED: Mites

RATES: Applied at 2-8 actual/100 gal of water or 1/2-2 lb actual/A.

APPLICATION: Apply when mites first appear, and repeat as necessary.

PRECAUTIONS: Agitation is required while spraying. Thorough coverage is necessary for good control. Toxic to fish. May not be compatible with Bordeaux mixture. Do not add oil to the spray solution.

ADDITIONAL INFORMATION: Not highly injurious to beneficial mites, and Non-toxic to honey bees. Nonvolatile and nonsystemic, so it kills by contact action. No ovicidal activity. Mites are paralyzed prior to death and sometimes take 2-3 days to die. Gives long-lasting control. May be mixed with other pesticides. Safe to bees.

NAMES

***AZOCYCLOTIN*, BAY-BUE-1452, CLEARMITE, CLERMAIT, PEROPAL**

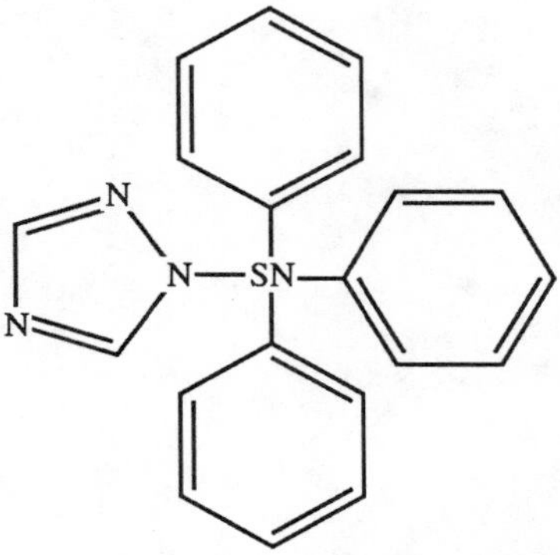

N
N—SN—
N

1-(Tricyclohexylstannyl)-1*H*-1,2,4-triazole

TYPE: Azocycloten is a heterocyclic tin compound being used as a contact acaracide.

ORIGIN: Bayer AG of Germany, 1980.

TOXICITY: LD_{50}-90 mg/kg. May cause eye and skin irritation.

FORMULATIONS: 50% WP, 25% WP.

PHYTOTOXICITY: Non-phytotoxic when used as directed.

IMPORTANT PESTS CONTROLLED: Mites.

USES: Experimentally being tested on vegetables, grapes, fruits, and citrus outside the U.S. Being sold in some countries for use on these crops.

RATES: Used as a .2-.3% concentration.

APPLICATION: Apply when mites appear and repeat as necessary.

PRECAUTIONS: Not for sale or use in the U.S. Toxic to fish.

ADDITIONAL INFORMATION: Harmless to bees. Controls both the larvae and adult mite stages. May have ovicidal activity. Does not have insecticidal activity. Long lasting on the plant. Compatible with other pesticides.

CARBAMATES

NAMES

***CARBARYL*, CARPOLIN, DENAPON, DICARBAM, HEXAVIN, KARBASPRAY, MURVIN, PATRIN, RAVYON, SEPTENE, SEVIN, TERCYL, TRICARNAM**

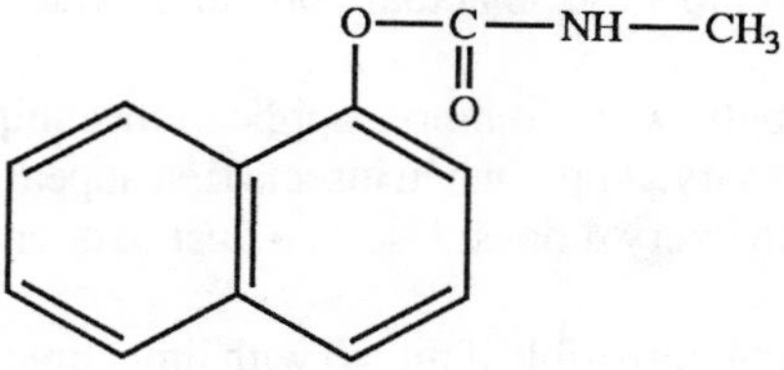

1-Naphthyl methylcarbamate

TYPE: Sevin is a carbamate insecticide, expressing contact and stomach-poison action with long residual effects.

ORIGIN: Rhone Poulenc. 1956.

TOXICITY: LD_{50}-500 mg/kg.

FORMULATIONS: 5 and 10% dusts, 5% and 10% granules, 4 lb/gal flowable, 50% and 80% WP.

PHYTOTOXICITY: Excessive dosages may retard germination of grasses. Injury may occur on tender foliage in the event of several days of rain or high humidity. Injury has been reported on McIntosh and York varieties of apple, and some pears. Watermelons and Boston ivy have been injured.

USES: Alfalfa, almonds, apples, apricots, asparagus, avocados, bananas, beans, beets, blueberries, broccoli, Brussels sprouts, cabbage, caneberries, carrots, cauliflower, celery, cherries, chestnuts, citrus, clover, collards, corn, cotton, cowpeas, cranberries, cucumbers, dandelion, dewberries, eggplant, endive, filberts, grapes, grasses, horseradish, kale, hanover salad, kohlrabi, lettuce, maple trees, melons, mustard greens, nectarines, okra, pastures, olives, parsley, parsnips, peaches, peas, peanuts, pears, peppers, pineapples, plums, potatoes, prunes, pumpkins, quince, radishes, rice, rutabagas, salisfy, sorghum, soybeans, spinach, squash, strawberries, sugar beets, sweet potatoes, Swiss chard, tobacco, tomatoes, trefoil, turnips, walnuts, horses, dogs, cats, poultry, agricultural premises, ornamentals, forest lands, wooded areas, rangeland, noncrop areas, and commercial buildings. Used outside the U.S. on these as well as many other crops.

IMPORTANT PESTS CONTROLLED: Aphids, codling moths, plum, curculio, leafhoppers, scale, bollworms, armyworms, pear psylla, lygus bugs, Japanese beetles, boll weevils, peach twig borers, spittlebugs, thrips, grasshoppers, stinkbugs, cucumber beetles, and many others, including ticks, fleas, and mites on dogs, cats, and poultry.

RATES: Applies at 1/2 to 1-1/2 lb actual/100 gal of water or 1/2 to 4 lb actual/A.

APPLICATION: Apply with common application equipment at a uniform rate. Repeat as necessary. Apply when insects first appear. Do not apply to livestock more often than every 4 days. Use as a dust bath on poultry, cats, and dogs.

PRECAUTIONS: Noncompatible if mixed with lime, lime sulfur, bordeaux, or other aklaline materials. Does not control spider mites. Highly toxic to bees. Do not use on apples as an insecticide at blossom time if they are to be chemically thinned with other compounds, or overthinning may occur.

ADDITIONAL INFORMATION: A very safe insecticide. No off-flavor has resulted on harvested crops. Compatible with insecticides and fungicides. Some systemic action has been shown. Controls the eggs of some insect species. Flies are tolerant to this compound. Toxicity increases as the temperature increases. Also used to chemically thin apples. Control should last for 1-3 weeks.

NAMES

***AMINOCARB*, BAY-44646, METACIL**

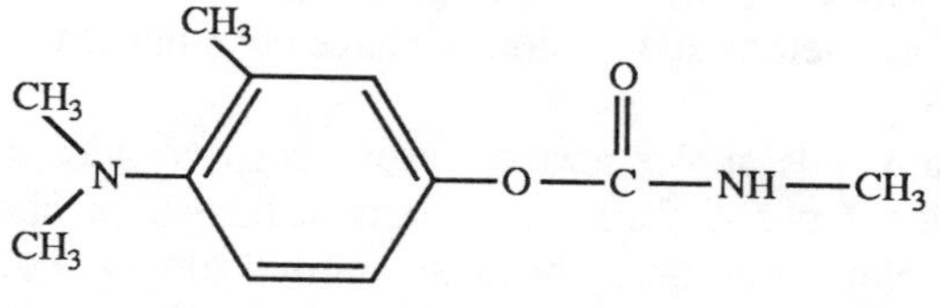

(4-Dimethylaminophenyl-3-methyl-phenyl)- N-methylcarbamate

TYPE: Metacil is a carbamate insecticide acting as a stomach and contact poison.

ORIGIN: Bayer AG in Germany, 1963.

TOXICITY: LD_{50}-30 mg/kg.

FORMULATIONS: 50 & 75% WP.

PHYTOTOXICITY: Nonphytotoxic at the recommended rates.

USES: Forest and ornamental trees outside the U.S.

IMPORTANT PESTS CONTROLLED: Spruce budworm and jackpine budworm. (Lepidoptera insects)

RATES: Applied at 1-16 oz ai/A.

APPLICATION: Thorough coverage is necessary. Spray when insects first appear and repeat as necessary.

PRECAUTIONS: Hazardous to bees. Do not combine with alkaline compounds. Not for sale in the U.S.

ADDITIONAL INFORMATION: Most effective against biting insects. Somewhat effective on mites. Compatible with most other pesticides.

NAMES

METHIOCARB, MERCAPTODIMETHUR, CLUB, DRAZA, MESUROL, METMERCAPTURON

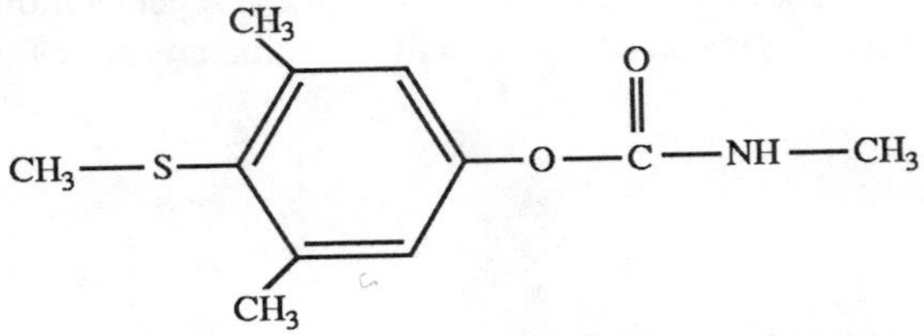

CH_3—S—(ring with two CH_3)—O—C(=O)—NH—CH_3

4-methylthio-3,5-xylylmethylcarbamate

TYPE: Mesurol is a carbamate insecticide-acaricide-molluscicide that kills by contact and stomach-poison action. Also used as a bird repellent.

ORIGIN: Bayer AG in Germany, 1962. Sold in the U.S. by Mobay Chemical Corp.

TOXICITY: LD_{50}-130 mg/kg.

FORMULATIONS: 50% and 75% WP, 2% bait, 3% dust, 4% pellets.

PHYTOTOXICITY: Fruit thinning may occur if applied earlier than 4 weeks after petal fall and before flowering.

USES: Turf, blueberries, cherries, citrus, corn, grapes, peaches, nonbearing fruit trees, and ornamentals. Being used in Europe on orchards, cotton, grapes, corn, apples, pears, ornamentals, hops, and vegetable crops. Being used experimentally in the U.S. on these and other crops. Also being developed for forest pest control. Being used as a bird repellent on some crops. Being used worldwide for slug and snail control.

IMPORTANT PESTS CONTROLLED: Thrips, slugs, snails, blueberry maggot, grasshoppers, fruit flies, plum curculio, pear psylla, mosquitoes, mites, leafhoppers, flies, aphids, codling moths, and many others.

RATES: Applied at 1/2-3 lb actual/A.

APPLICATION: Apply when insects appear and repeat as necessary. Lightly water the area prior to scattering the bait for slug and snail control. Being used experimentally as a seed treatment.

PRECAUTIONS: To avoid blossom-thinning of fruit trees, spray either before bloom, or wait until 6 weeks after blooming. Harmful to bees. Spray immediately after mixing with alkaline compounds. Toxic to fish. Somewhat toxic to earthworms. Do not mix with alkaline compounds.

ADDITIONAL INFORMATION: Kills predominately by contact activity. Fast cleanup with a long residual activity. Doesn't penetrate the plant. Compatible with most other pesticides. Repellency to birds has been demonstrated on corn and fruit crops, and is used commercially in some areas for this purpose.

NAMES

***TRIMETHACARB,* BROOT, LANDRIN, SD 8530, UC 27867**

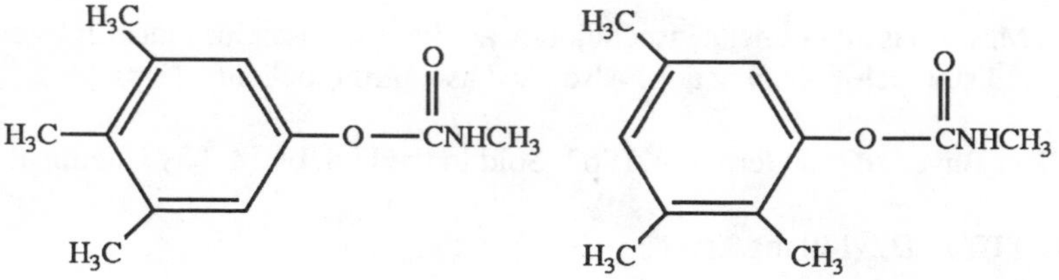

Trimethylphenyl methylcarbamate

TYPE: Trimethacarb is a carbamate insecticide, acting primarily by ingestion, with some contact activity.

ORIGIN: Shell Chemical Co., 1964. Purchased by Rhone Poulenc, 1982. Now being sold by Drexel Chemical Co.

TOXICITY: LD_{50}-130 mg/kg.

FORMULATIONS: 15 G, 50 WP, 75 WP, 4 lb/gal flowable.

PHYTOTOXICITY: No phytotoxicity indicated when applied as a foliar spray to a wide variety of agricultural and ornamental plants. Phytotoxicity has been noted in some crops (eg., corn, sorghum, wheat, rice) where direct contact of chemical and seed occurs.

USES: Currently registered on corn. Experimentally being tested on a wide variety of crops. Activity is also indicated as a molluscicide, as a mammal and avian aversion agent, and as a control measure for pests of public health importance.

IMPORTANT PEST CONTROLLED: Corn rootworm larvae.

RATES: Apply granules at 1 lb a.i./A in a 6-8 inch band over the raw.

APPLICATION: Incorporate the granules of applying behind the covering disk and ahead of the press wheel.

PRECAUTIONS: Do not harvest within 90 days of application. Toxic to fish.

ADDITIONAL ACTIVITY: Season long residual activity. Noncorrosive.

NAMES

ISOPROCARB, BAY 105807, ETROFOLAN, HYTOX, MIPC, MIPCIN

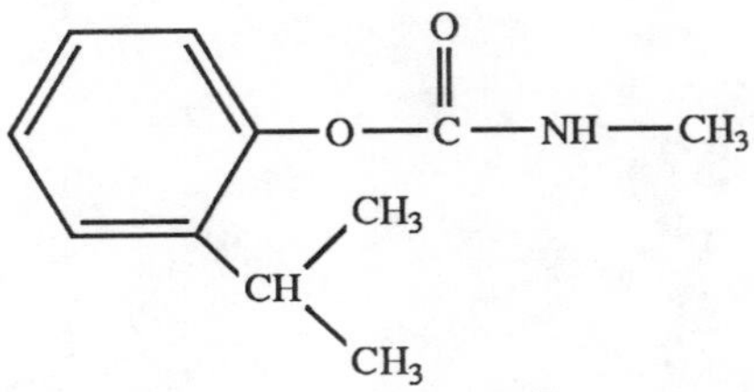

2-Isopropyl-phenyl-*N*-methylcarbamate

TYPE: Etrofolan is a carbamate insecticide, controlling insects as a stomach-poison.

ORIGIN: Bayer AG of Germany, 1972.

TOXICITY: LD_{50}-178 mg/kg.

FORMULATIONS: 15% thermal fog, 5% dust, 5% granules, 50% WP, EC 200 g/1.

PHYTOTOXICITY: Nonphytotoxic when used as directed.

USES: Outside the U.S. on cocoa and rice. Being developed for use on cotton, tree crops, peanuts, vegetables, cereals, hops, coffee, potatoes, orchards, sugarcane, and ornamentals. Being developed in the U.S. for use on alfalfa, pastures, and cereals.

RATES: Applied at 250-1000 a.i./ha.

IMPORTANT PESTS CONTROLLED: Plant hoppers, leafhoppers, aphids, plant bugs, thrips, stinkbugs, codling moth, psylla, and many others.

APPLICATION: Apply when insects appear and repeat as necessary.

PRECAUTIONS: Harmful to bees. Do not use on rice 10 days prior to, or after, a propanil application. Do not combine with alkaline pesticides.

ADDITIONAL INFORMATION: May be combined with other pesticides. Fast acting, with moderately long residual activity.

RELATED COMPOUNDS:

1. HOPCIDE, ETHROFOL, CPMC — A carbamate insecticide developed by Bayer AG & Kumiai Chemical Company of Japan to control leafhoppers and planthoppers on rice.

NAMES

PROPOXUR, APROCARB, BAYGON, BLATTANEX, PHC, PROPION UNDEN, SENDRAN, SUNCIDE

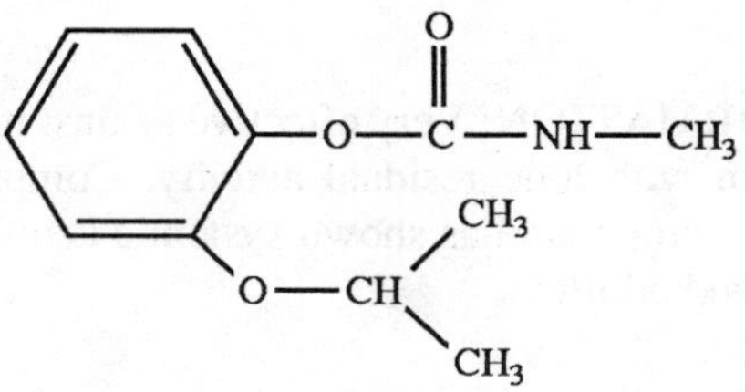

2-(1-Methylethoxy)phenyl methylcarbamate

TYPE: Baygon is a carbamate insecticide with contact and stomach-poison activity.

ORIGIN: Bayer AG in Germany, 1959. Being developed in the U.S. by Mobay Chem. Corp.

TOXICITY: LD_{50}-90 mg/kg.

FORMULATIONS: 70% WP, 1.5 EC, 1.4 oil-soluble formulation/gal, EC-200 g/1, 50% WP 2% dust. Aerosols.

PHYTOTOXICITY: Some injury has been reported on chrysanthemums, carnations, and hydrangeas at higher rates. May cause blossom-thinning of fruit trees.

USES: Structural pest control, domestic pets, ornamentals, turf, mosquito control, agricultural premises, commercial premises, and in households. Being used in many countries on alfalfa, corn, soybeans, orchards, vineyards, vegetables, potatoes, cotton, sugarcane, forests, rice, cocoa, and other crop plants.

IMPORTANT PESTS CONTROLLED: Earwigs, crickets, chinch bugs, cockroaches, ants, grasshoppers, aphids, leafhoppers, flies, fleas, thrips, mites, mosquitoes, spiders, ticks, and many others.

RATES: Applied at 1/8 to 1 lb a.i./A or at .5-2% concentrations for household sprays.

APPLICATION: Apply evenly and thoroughly, repeating when necessary. Such substances as wood, tiles, paints, clay, etc., may be treated where insects occur. For severe infestations, repeat application in four to six weeks. Bait formulations are used to control crawling insects in structures. Used both indoors and outside. Applied to turf for sod webworms and chinch bug control.

PRECAUTIONS: Use with water at below 40°F, or with oil, results in crystallization in the spray tank. Do not use as a space spray. Harmful to bees. Spray immediately after mixing with alkaline materials. Do not use fiberglass tanks. Birds feeding on treated areas may be killed. Toxic to fish. Do not apply to tidal marshes or estuaries. Do not spray on orchards earlier than 3 weeks after petal fall.

ADDITIONAL INFORMATION: Very effective against most public health pests. Fast knockdown with long residual activity. Compatible with most other pesticides. This compound has shown systemic activity when applied to the soil. Colorless and odorless.

NAMES

***METOLCARB*, METACRATE, MTMC, TSUMACIDE**

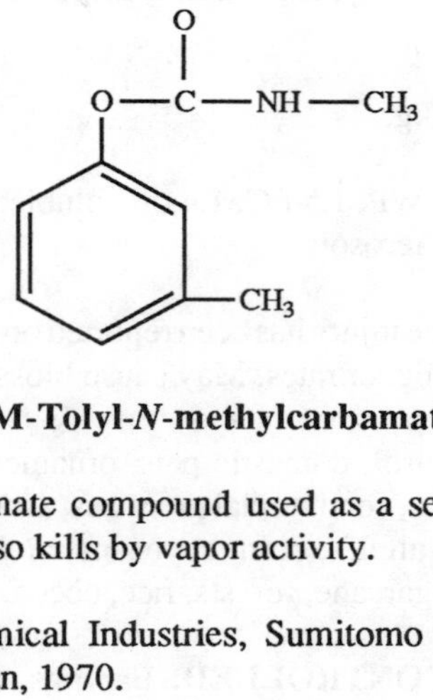

O
|
O—C—NH—CH_3

CH_3

3-M-Tolyl-*N*-methylcarbamate

TYPE: MTMC is a carbamate compound used as a selective, contract, systemic insecticide, which also kills by vapor activity.

ORIGIN: Mitsubishi Chemical Industries, Sumitomo Chemical Co. and Nihon Nohyaku Co. of Japan, 1970.

TOXICITY: LD_{50}-498 mg/kg.

FORMULATIONS: 30% EC, 50% WP, 2% dust.

PHYTOTOXICITY: Nonphytotoxic.

IMPORTANT PESTS CONTROLLED: Plant hoppers and leafhoppers.

USES: Paddy-rice plants in Japan.

APPLICATION: Apply when insects appear and repeat as necessary, usually at 10-13 day intervals.

PRECAUTIONS: Not for sale or use in the U.S. Do not mix with alkaline products such as Bordeaux.

ADDITIONAL INFORMATION: Compatible with other pesticides. Quick-acting. Does not harm spiders which prey on leafhoppers. Low toxicity to fish. Effective at low temperatures.

RELATED MIXTURES:

1. TUMAZINON—A combination insecticide of 3% diazinon and 2% MTMC sold in Japan by Mitsui Toatsu Chemicals, Inc. for use on rice.

NAMES

CLOETHOCARB, BAS-263, LANCE

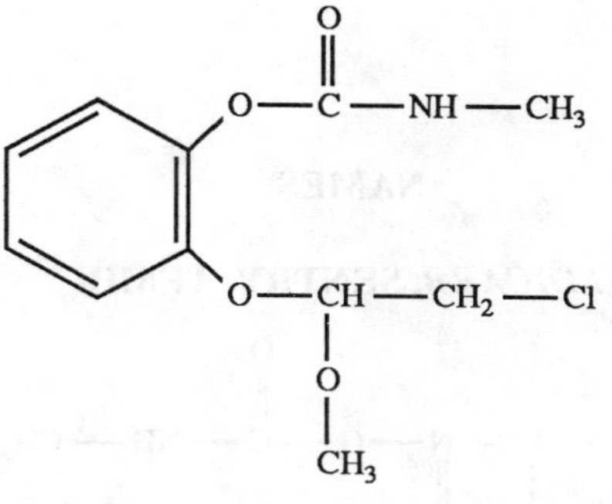

1-(2-Chloro-1-methoxyethoxy)phenyl N-methylcarbamate

TYPE: Cloethocarb is a carbamate compound being used as a broad spectrum contact stomach and systemic nematicide.

ORIGIN: BASF of Germany, 1978.

TOXICITY: LD_{50} -35 mg/kg.

FORMULATIONS: 50% WP, 5% and 15% granules, 480 g/1 liquid seed treatment.

PHYTOTOXICITY: None observed when used as directed.

USES: Experimentally being tested on corn, rapeseed, potatoes, soybeans, rice, cereals, and other field crops, vegetables, and tree crops.

IMPORTANT PESTS CONTROLLED: Corn rootworm, nematodes, Colorado potato beetle, aphids, scales, pear psylla, caterpillars, and others.

RATES: Applied at 0.3-2 kg a.i./ha as a soil insecticide-nematicide; as a seed treatment applied at 0.15-0.5 kg ai/100 kg of seed; as a foliar treatment at 0.25-0.75 kg a.i./ha.

APPLICATION: Applied to corn in a 7 inch band ahead of the planter press wheel or directly in the seed furrow with the seed. Applied to rapeseed either as a mixture of granules and seed, with a hoe, press or disc drill, or with common granule application equipment. May also be applied as a band treatment at planting (on potatoes, soybeans and other field crops), as seed treatment on cereals, or as foliar spray.

PRECAUTIONS: Used on an experimental basis only. Moderately toxic to fish.

ADDITIONAL INFORMATION: Very effective against soil insects. Absorbed by the root system and translocated to the foliage. Residual in the range of 3-7 weeks.

NAMES

ALDICARB, SENTRY, TEMIK

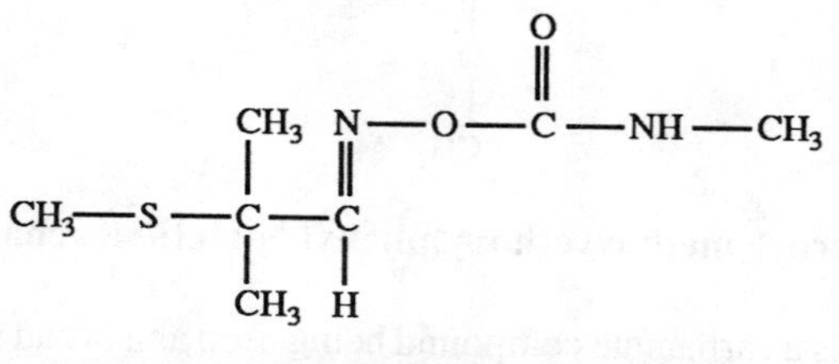

```
                       O
                       ||
           CH3  N──O──C──NH──CH3
            |   ||
CH3──S──C──C
            |   |
           CH3  H
```

2-Methyl-2-(methylthio)propionaldehyde
O-(methylcarbamoyl) oxime

TYPE: TEMICK is a carbamate compound used as a systemic insecticide-acaricide-nematocide.

ORIGIN: Union Carbide Corporation, 1965. Now produced by Rhone Poulenc.

TOXICITY: LD_{50}-.9 mg/kg. Rapidly absorbed through the skin. Very high toxicity.

FORMULATIONS: 10% and 15% granules.

PHYTOTOXICITY: Not applied to plant foliage. Seed germination is not inhibited at use rates. When applied in the furrow with acid-delinted cottonseed during extremely wet weather conditions, the plant stand may be slightly reduced.

USES: Cotton, potatoes, sweet potatoes, pecans, tobacco, seed alfalfa, dry beans, citrus, sorghum, soybeans, coffee, sugarcane, sugar beets, peanuts, and ornamentals.

IMPORTANT PESTS CONTROLLED: Aphids, mites, Colorado potato beetles, thrips, lygus, fleahoppers, boll weevils, fleabeetles, wireworms, leafminers, webworms, mealy bugs, leafhoppers, nematodes, and many others.

RATES: Applied at 1/2-10 lb actual/A.

APPLICATION: Applied as an in-furrow treatment at planting time. Also broadcast and side-dress treatments may be utilized. To be effective, the chemical must reach the root zone. Watering after application will improve the effectiveness. Cover with soil immediately after application to protect wildlife. Do not place less than 2 inches of soil on top of the granules.

PRECAUTIONS: Very highly toxic material. Lepidopterous insects, such as cutworms, armyworms, loopers, bollworms, and cornborers are not controlled effectively. Noncompatible with alkaline compounds. Do not handle with bare hands. Do not apply by aircraft. Do not use in or around homes or home gardens. Do not graze treated areas. Should be applied on ornamentals only by trained professionals.

ADDITIONAL INFORMATION: Residual control of up to 10 weeks can be expected, due to systemic activity. Very effective on soil nematodes, such as the spiral, ring, leison, dagger, root-knot, and stubby species. Due to the toxicity, the granules are covered with a water-soluble coating. Relatively Non-toxic to fish. Soil type or pH does not affect the activity. Systemic activity is upward only in the plant. Fast acting, with effects noticeable in 48 hours. Effective in all types of soil.

NAMES

PIRIMICARB, ABOL, AFICIDA, APHOX, FERNOS, PIRIMOR, RAPID

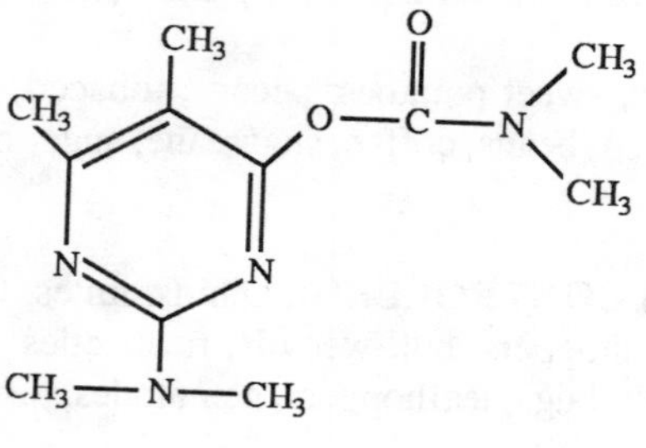

5,6-Dimethyl-2-dimethylamino-4-pyrimidinyl-dimethylcarbamate

TYPE: Pirimor is a carbamate compound used as a selective insecticide. Effective on contact, and by fumigation activity.

ORIGIN: ICI of England (Plant Protection, Ltd.), 1969.

TOXICITY: LD_{50}-147 mg/kg.

FORMULATIONS: 50% WP, 5% EC, .1% aerosol.

PHYTOTOXICITY: Nonphytotoxic when used at the recommended rates.

USES: No longer being sold in the U.S. Being used on many vegetable, ornamental, field, and fruit crops outside the U.S.

IMPORTANT PESTS CONTROLLED: Aphids.

RATES: Being used at a 2-4 oz a.i./A.

APPLICATION: Apply when insects appear and repeat as necessary. May be applied by air.

PRECAUTIONS: Not for sale or use in the U.S.

ADDITIONAL INFORMATION: Fast acting. Relatively short residual. Very selective. Does not harm lady bugs or lacewings. Not translocated extensively from foliar applications. Has a quick knockdown effect. Ideal for use in greenhouses. Low bee toxicity, so it may be applied to the open chrysanthemum flower. May be applied with other pesticides.

NAMES

***CARBOFURAN*, CARBODAN, CARBOSIP, CHINUFUR, CURATERR, FURADAN, KENOFURAN, YALTOX**

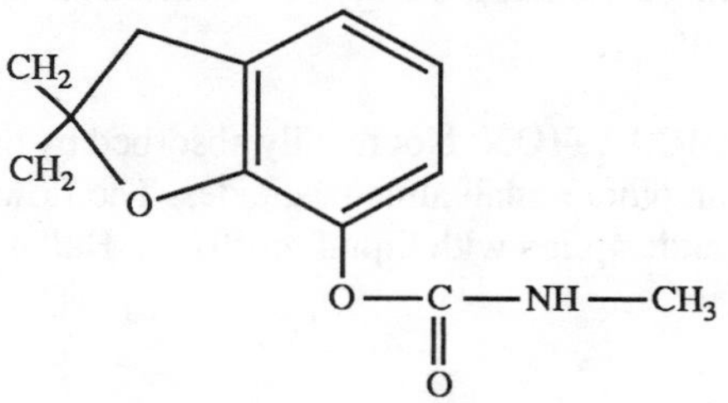

CH_2

CH_2 O

O—C—NH—CH_3

O

2,3-Dihydro-2,2-dimethyl-7-benzofuranyl methylcarbamate

TYPE: Carbofuran is a carbamate compound used as a systemic and contact stomach-poison insecticide-nematocide

ORIGIN: FMC Corp. and Mobay Chem. Corp., 1969

TOXICITY: LD_{50}-8 mg/kg.

FORMULATIONS: 2, 3, 5, and 10% granules, 4 lb/gal flowable, 330 g/1 suspension.

PHYTOTOXICITY: Nonphytotoxic when used as directed.

USES: Corn, cotton, cucurbits, small grains, grapes, alfalfa, artichokes, peanuts, peppers, strawberries, pines, soybeans, sunflowers, tobacco, bananas, sorghum, potatoes, cottonwood trees, sugarcane, and rice. Used outside the U.S. on a number of crops such as bananas, coffee, rice, sugar beets, pumpkins, squash, cucumbers, melons, and grapes.

IMPORTANT PESTS CONTROLLED: Corn rootworms, wireworms, boll weevils, sugarcane borer, thrips, mosquitoes, nematodes, rice water weevil, alfalfa weevil, aphids, scale, European corn borer, armyworms, and others.

RATES: Applied at 1/4-1 lb actual/A on foliage, 1/2-3 lb/A in the seed furrow and 6-10 lb/A for nematode control.

APPLICATION: Applied to corn in a 7 inch band or in furrow at planting time. Incorporate into the top 1 inch of soil by applying ahead of the press wheels or by using special covering devices. Apply on cane over the planted cane, just prior to covering with soil. On rice, use prior to flooding, or within 21 days after flooding. On peanuts and peppers, apply and incorporate into the soil prior to planting. May also be applied to the corn foliage and to alfalfa foliage.

PRECAUTIONS: Toxic to fish and wildlife. Will support combustion if ignited. Do not mix with alkaline pesticides. Do not apply to rice within 21 days of a propanil herbicide application, nor apply propanil afterwards. Avoid drift. Toxic to bees. Exposed granules at the end of the rows should be disked into the soil. Birds feeding in treated areas my be killed. Do not apply through overhead sprinkler systems.

ADDITIONAL INFORMATION: Not readily absorbed by the skin. Noncorrosive. Compatible with other nonalkaline pesticides. The flowable formulation may be tank mixed and applied with liquid fertilizers. Half-life in the soil is 30-60-days.

NAMES

CARBOSULFAN, ADVANTAGE, MARSHALL, POSSE

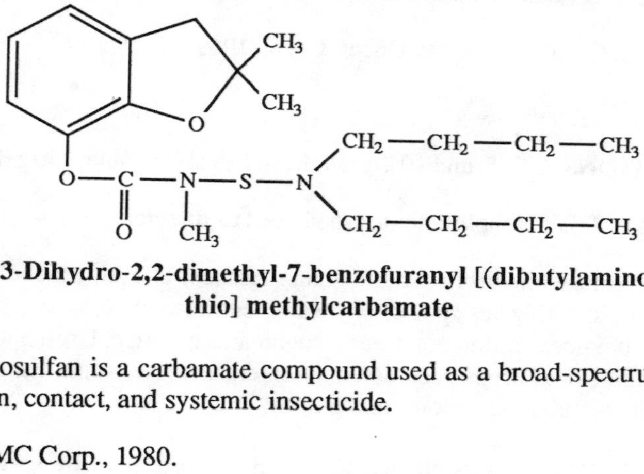

2,3-Dihydro-2,2-dimethyl-7-benzofuranyl [(dibutylamino) thio] methylcarbamate

TYPE: Carbosulfan is a carbamate compound used as a broad-spectrum stomach-poison, contact, and systemic insecticide.

ORIGIN: FMC Corp., 1980.

TOXICITY: LD_{50}-209 mg/kg. May cause skin and eye irritation.

FORMULATIONS: 2.5 EC, 4 EC, 5 G, and 15 G.

PHYTOTOXICITY: Non-phytotoxic when used as directed.

USES: Being used outside the U.S. on apples, alfalfa, corn, citrus, sorghum, soybeans, and other crops.

IMPORTANT PESTS CONTROLLED: Weevils, aphids, lygus, leafhoppers, grasshoppers, codling moth, scale, corn rootworm, wireworm, European corn borer, nematodes, mites, thrips, greenbug, chinch bug, and others.

RATES: Applied at .5-2 lb a.i./A.

APPLICATION: Apply to the soil at planting, and lightly incorporate. Also applied as a foliar spray.

PRECAUTIONS: Not for sale in the U.S. as yet.

ADDITIONAL INFORMATION: May be mixed with liquid fertilizer. Seed treatments are under investigation.

NAMES

***ALDOXYCARB*, STANDAK, SULFOCARB, UC-21865**

$$CH_3-\overset{O}{\underset{O}{\overset{\|}{\underset{\|}{S}}}}-\overset{CH_3}{\underset{CH_3}{\overset{|}{\underset{|}{C}}}}-CH=N-O-\overset{O}{\overset{\|}{C}}-NH-CH_3$$

2-Methyl-2-(methyl sulfonyl) propanal *O*-(methylamino) carbonyl oxime

TYPE: Aldoxycarb is a carbamate compound used as a systemic insecticide-nematicide.

ORIGIN: Union Carbide Corp., 1976. Now being developed by Rhone Poulenc.

TOXICITY: LD_{50}-21.4 mg/kg. May cause slight skin irritation.

FORMULATIONS: 2.67 lb/gal flowable, 5% granules.

PHYTOTOXICITY: Injury has been observed on soybeans and snap bean seedlings.

USES: Experimentally being tested on tobacco, cotton, peanuts, vegetables, and others.

IMPORTANT PESTS CONTROLLED: Nematodes, thrips, aphids, mites, Colorado potato beetle, Mexican bean beetle, flea beetle, plant bugs, leafhoppers, and others.

RATES: As a soil treatment applied at 2-3 lbs a.i./A. Applied as a transplant water treatment at 1.5-2 lb a.i./A.

APPLICATION: Applied as a soil treatment, either as a broadcast application and incorporated, or as a banded in-furrow treatment prior to planting. Applied in transplant water at time of transplanting. Seed treatments are under investigation.

PRECAUTIONS: Used on an experimental basis only. Not effective on lepidopterous larvae or soil insects.

ADDITIONAL INFORMATION: A nonfumigant nematicide. Absorbed by the roots and translocated to the foliar portions of the plant. Residual activity in the range of 4-8 weeks. Low toxicity to bees.

NAMES

***THIODICARB*, LARVIN, MAGNUM, NIVRAL, SEMEVIN, SKIPPER, UC 51762, UC-80502**

$$CH_3-C(SCH_3)=N-O-C(=O)-N(CH_3)-S-N(CH_3)-C(=O)-O-N=C(SCH_3)-CH_3$$

Dimethyl N,N'(thiobis((methylimino) carbonoyloxy)) bisethanimidothioate

TYPE: LARVIN is an oxime carbamate insecticide being used as a stomach poison insecticide but does not have limited contact toxicity.

ORIGIN: Union Carbide, 1978. Union Carbide Corp. 1978. Now being developed by Rhone Poulenc.

TOXICITY: LD_{50}-66 mg/kg.

FORMULATIONS: 250 g/l, 375 g/l, 3.2 lb/gal, 80% dry flowable, 75% wettable powder, 5% bait/granules, 2% dust. 3.2 lb/gal. flowable.

PHYTOTOXICITY: Non-phytotoxic when used as recommended.

USES: EPA registered on fresh market sweet corn, cotton and soybeans. Experimentally being tested on, vegetables, corn, and many other crops.

IMPORTANT PESTS CONTROLLED: Lepidopterous pests such as armyworms, budworms, bollworms, leaf rollers, cutworms, corn earworms, leafworms, and loopers. Also controls some coleoptera, Diptera boll weevil, European corn borer, and Homoptera pests of agricultural crops and forests.

RATES: Applied at .25-.9 a.i./acre.

APPLICATION: Begin application when insect populations reach recognized economic threshold levels. Repeat application as needed to maintain control. Use high dosage rates for heavier infestation or larger larvae.

PRECAUTIONS: Do not combine with heavy metal fungicides such as maneb, mancozeb, Copper Count-N or Bordeaux. Thiodicarb is hydrolytically sensitive to degradation of active ingredient by strong acids, strong bases, and certain heavy metal oxides and salts and the metal/salt complexes of certain fungicides. Do not add thiodicarb to water with pH values below 3.0 or above 8.5. The product is not miscible with vegetable oil diluents. Prepared spray mixtures should be used within 6 hours after mixing.

ADDITIONAL INFORMATION: Moderate knockdown and good residual control of both the larvae and adult insects. Gives 5-14 days residual activity. Some ovicidal and adulticidal activity on Lepidoptera insects. Has limited systemic activity as a seed treatment. Active on slugs when used as a bait.

NAMES

***ALANYCARB*, OK-135, ORION**

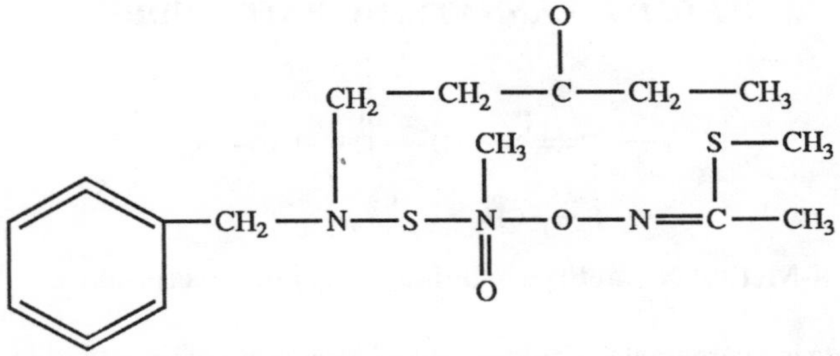

S-Methyl-N-[[N-methyl-N-[N-benzyl-N-(2-ethoxy-carbonylethyl) aminothio]carbamoyl]thioacetimidate

TYPE: Alanycarb is a carbamate compound used as a stomach poison and contact insecticide.

ORIGIN: Otsuka Chemical Co. of Japan, 1982.

TOXICITY: LD_{50}-301 mg/kg. May cause skin and eye irritation.

FORMULATION: 30% EC, 40% WP, 3% G, 5% G.

PHYTOTOXICITY: Non-phytotoxic when used as directed.

USES: Experimentally being tested on fruit trees, corn, cotton, grapes, vegetables, peanuts, potatoes, soybeans, sugar beets, tea, tobacco, and others.

IMPORTANT PESTS CONTROLLED: Aphids, codling moth, leafrollers, armyworms, bollworm, budworm, leafhoppers, thrips, corn earworm, potato tuberworm, cutworms, flea beetles, cabbage looper, cabbageworm, hornworms and others.

RATES: Applied at 300-600 g a.i./ha.

APPLICATION: Apply as a foliar spray. Repeat as needed. The granules are applied as a soil treatment.

PRECAUTIONS: Not for sale or use in the U.S. Toxic to fish.

ADDITIONAL INFORMATION: Most effective on lepidoptera insects. Excellent residual activity.

NAMES

METHOMYL, LANNATE, NU-BAIT, NUDRIN

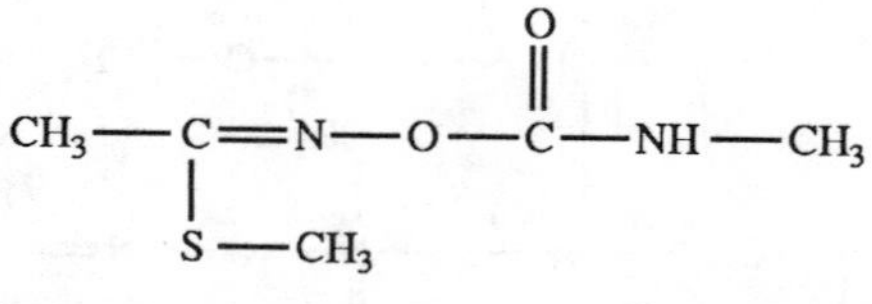

$$CH_3-\underset{\underset{S-CH_3}{|}}{C}=N-O-\overset{\overset{O}{\|}}{C}-NH-CH_3$$

S-Methyl N-(methylcarbamoyl)oxy) thioacetimidate

TYPE: Lannate is a carbamate compound, used as a systemic insecticide-nematocide, killing by contact as a stomach-poison.

ORIGIN: E. I. DuPont de Nemours and Co. 1967.

TOXICITY: LD_{50}-17 mg/kg.

FORMULATIONS: 90% water soluble powder, 1.8 lb/gal water-soluble liquid.

PHYTOTOXICITY: Do not use on early MacIntosh or Wealthy varieties of apples. Do not apply to spinach when daily temperature is 32°F or under, or to seedlings less than 3 inches in diameter. Redding of cotton may occur.

USES: Alfalfa, apples, artichokes, asparagus, avocados, beans, beets, blueberries, broccoli, Brussels sprouts, cabbage, carrots, cauliflower, celery, Chinese cabbage, citrus, collards, cotton, corn, cucumbers, chickory, dandelions, eggplant, endive, escarole, grapes, hops, horseradish, kale, lentils, lettuce, melons, mustard, nectarines, onions, ornamentals, parsley, peanuts, peas, pecans, peppers, pomegranates, potatoes, radishes, rutabagas, sorghum, Swiss chard, turnips, watercress, soybeans, spinach, squash, sugar beets, tobacco, tomatoes, and turf.

IMPORTANT PESTS CONTROLLED: Loopers, bollworm, alfalfa weevils, European cornborer, leafhoppers, cabbageworms, tobacco budworms, flea beetles, tomato fruitworm, aphids, tobacco hornworm, corn earworm, armyworms, and many others.

RATES: Applied at .225-.9 a.i./A.

APPLICATION: Apply when insects appear and repeat as necessary, usually at 5-7 day intervals. May be applied with a wetting agent.

PRECAUTIONS: Toxic to bees.

ADDITIONAL INFORMATION: Insects fall to the ground within minutes of contacting the spray. Systemic in activity. Compatible with other pesticides.

NAMES

OXAMYL, THIOXAMYL, VYDATE

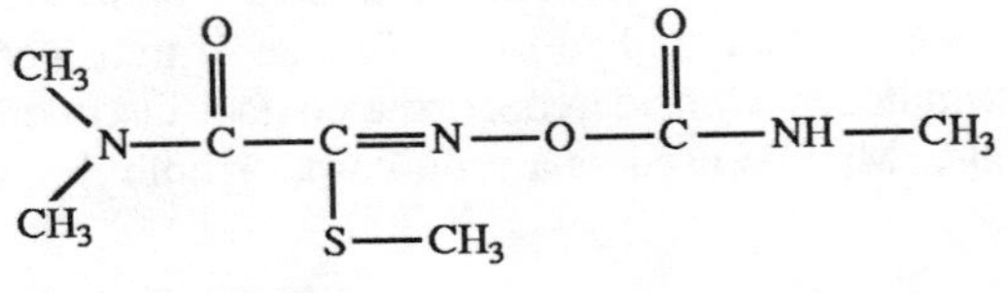

```
             O                    O
 CH3         ‖                    ‖
    \        ‖                    ‖
     N — C — C = N — O — C — NH — CH3
    /            |
 CH3             |
                 S — CH3
```

Methyl N',N'-dimethyl-N-((methylcarbamoyl)oxy)-1-oxamimdate

TYPE: Oxamyl is a carbamate compound, used as a contact and systemic insecticide and nematicide.

ORIGIN: E. I. DuPont de Nemours and Co., 1972.

TOXICITY: LD_{50}-5.4 mg/kg.

FORMULATIONS: 2 EC, 10% granules.

PHYTOTOXICITY: Non-phytotoxic when used as directed. Some strawberry varieties will be injured.

IMPORTANT PESTS CONTROLLED: Flea beetles, leaf miners, nematodes, scales, leafhoppers, thrips, aphids, Japanese beetle, and boll weevil.

RATES: Applied at 1/4-1 lb actual/A for foliage insects and 2-8 lb/A for soil insects and nematodes.

USES: Tobacco, peanuts, eggplant, bananas, mint, cotton, peppers, soybeans, citrus, celery, non-bearing fruit trees, apples, pineapples, and ornamentals. Experimentally being used as a foliar and soil insecticide on fruit crops, vegetables, potatoes, peanuts, and others.

APPLICATION:

1. Foliar-Apply when insects appear and repeat as necessary.
2. Transplant water treatment-Apply in the transplant water when plants are set out.
3. Soil-Apply as a preplant, soil incorporated treatment, in transplant water, and as an in-furrow application. Plants should be set out within 24 hours. Also used as a root, corn, and bulb dip, a liquid drench, and a soil-mix treatment.

PRECAUTIONS: Toxic to bees. Birds and wildlife feeding in the treated area may be killed. Toxic to fish. Not compatible with alkaline materials. Wireworms are not controlled.

ADDITIONAL INFORMATION: Ineffective on wireworms. Considered to have moderate residual effects. Considered to be systemic when applied as a soil treatment. Also, when applied to the foliage, it translocates downward and controls nematodes. Considered to be a contact insecticide when applied to plant foliage. May be used as a root, corn, or bulb dip. Can be used inside greenhouses.

NAMES

FORMETANATE, CARZOL, DICARZOL

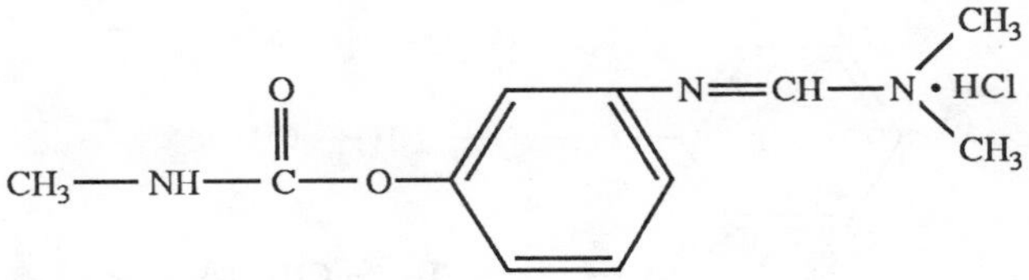

N,N-Dimethyl-N'-(3(((methylamino)carbonyl)oxy)phenyl) methanimidamide-monohydrochloride

TYPE: Formetanate is a carbamate compound being used as a contact acaricide and insecticide.

ORIGIN: Schering AG of Germany, 1967. Developed by NOR-AM Agricultural Products, Inc., in the U.S.

TOXICITY: LD_{50}-3.1 mg/kg. May cause eye irritation.

FORMULATIONS: 92% soluble powder and 50% WP.

PHYTOTOXICITY: Injury has been reported on peas, beans, peanuts, soybeans, eggplant, cucumbers, and certain rose varieties at high rates of application.

USES: Alfalfa, nectarines, plums, citrus, apples, pears, peaches, and prunes. Used outside the U.S. on a number of crops.

IMPORTANT PESTS CONTROLLED: Mites, thrips, leafhoppers, lygus, slugs, snails, stink bugs, fleabeetles, leaf miners, and many others.

RATES: Apply at 1-8 oz/100 gal or at 1/2-1 lb actual/A.

APPLICATION: Apply when mites or insects appear, and repeat as necessary. Thorough coverage is necessary.

PRECAUTIONS: Do not use in water above the pH of 8. Do not spray when rain is expected. Toxic to bees. Toxic to fish. Do not mix with alkaline pesticides. Do not allow livestock to graze treated areas. Do not prepare more spray than will be used within 4 hours. Aphids are not controlled.

ADDITIONAL INFORMATION: Effective on the mobile stages of mites. Compatible with most other pesticides.

NAMES

ETHIOFENCARB, BAY-HOX-1901, CRONETON

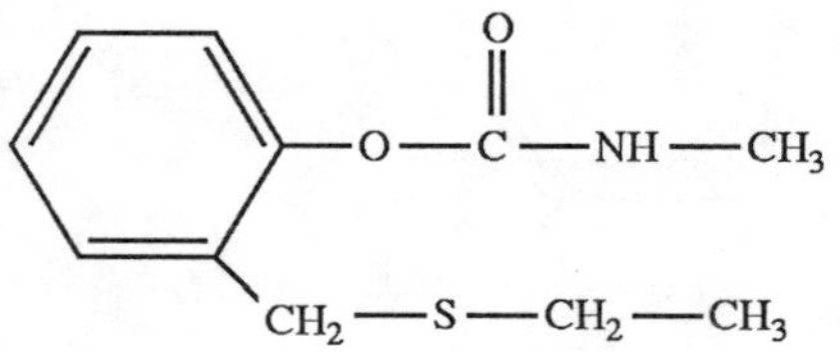

$$O$$

$$-O-C-NH-CH_3$$

$$CH_2-S-CH_2-CH_3$$

2-((Ethylthio)methyl)phenyl methylcarbamate

TYPE: Croneton is a carbamate compound, effective as a selective, stomach-poison, systemic, and contact insecticide.

ORIGIN: Bayer AG of Germany, 1975.

TOXICITY: LD_{50}-411 mg/kg.

FORMULATIONS: 4 EC, 10 & 15% granules, 40% WP, 500 g/1 EC.

PHYTOTOXICITY: Do not use on anthurium or begonias.

USES: Used on many fruit, field, ornamental and vegetable crops outside the U.S.

IMPORTANT PESTS CONTROLLED: Aphids.

RATES: Applied at 4-8 oz. a.i./A.

APPLICATIONS: Apply to the foliage when the insects appear, and repeat as necessary. Also applied as a soil drench or soil application.

PRECAUTIONS: Not for use in the U.S. Moderately toxic to fish.

ADDITIONAL INFORMATION: Good residual activity. Only effective on sucking insects. Does not harm bees. Compatible with other pesticides. May be systemic when applied to the root system.

NAMES

***FENOTHIOCARB*, KCO-3001, PANOCON, PANOSIN**

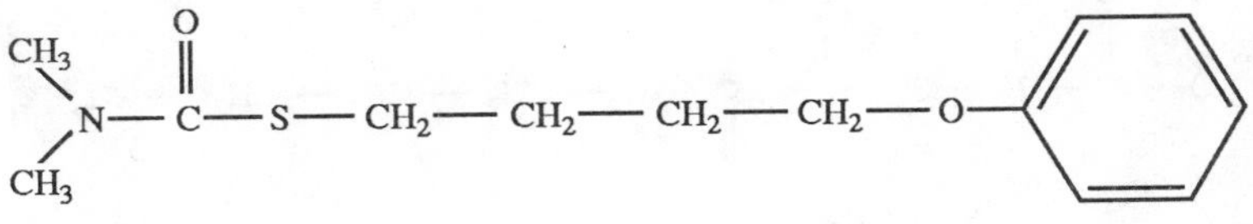

$$\begin{matrix} CH_3 & & O & & & & & & \\ & \diagdown & \| & & & & & & \\ & N - & C - & S - & CH_2 - & CH_2 - & CH_2 - & CH_2 - & O - C_6H_5 \\ & \diagup & & & & & & & \\ CH_3 & & & & & & & & \end{matrix}$$

S-(4-Phenoxybutyl)-N,N-dimethyl thiocarbamate

TYPE: Fenothiocarb is a carbamate compound used as an acaricide.

ORIGIN: Kumiai Chemical Ind. Co. of Japan, 1978.

TOXICITY: LD_{50}-1150 mg/kg.

FORMULATION: 35% EC. 65% EC.

PHYTOTOXICITY: Some injury at high rates has been reported on certain apple varieties, cotton, peaches, melons, and others.

USES: Used outside the U.S. on citrus, apples, cotton, vegetables, beans, pears, and other crops.

IMPORTANT INSECTS CONTROLLED: Mites

RATES: Applied at 350-500 ppm a.i.

APPLICATION: Apply when mites first appear and repeat as necessary.

PRECAUTIONS: Not for sale or use in the U.S. Do not mix with alkaline materials.

ADDITIONAL INFORMATION: No insecticidal activity. Especially effective against the mite genus PANONYCHUS. All mite stages are controlled, but especially effective against the eggs. Relatively Non-toxic to fish. Compatible with other pesticides. Long residual activity. Very active at low temperatures.

NAMES

FURATHIOCARB, CGA-73102, DELTANET, PROMET

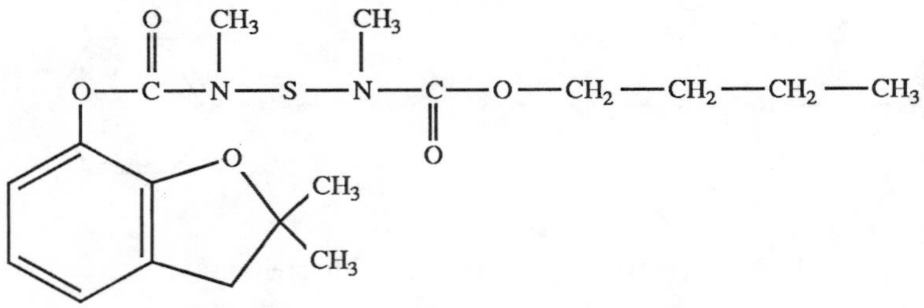

O-n-Butyl O'-(2,2-dimethyl-2,3-dihydro-benzofuran-7-yl)
N,N'-dimethyl N,N'-thiodicarbamate

TYPE: Furathiocarb is a carbamate compound used as a contact and stomach-poison insecticide-nematicide.

ORIGIN: CIBA-Geigy Corp., 1982.

TOXICITY: LD_{50}-100 mg/kg.

FORMULATION: Under evaluation.

PHYTOTOXICITY: Non-phytotoxic when used as directed.

USES: Experimentally being tested on corn, field, fruit, and vegetable crops. Being used on these crops outside the U.S.

IMPORTANT PESTS CONTROLLED: Corn rootworm, nematodes, scales, mites, aphids, and others. Some activity as a bird repellent.

RATES: Applied at 5-20 a.i./kg or 2-4 oz a.i./100 gal of water.

APPLICATION: Applied as an in-furrow treatment at planting time. Soil applications are made for the control of nematodes. Apply as a foliar spray when insects appear and repeat as necessary.

PRECAUTIONS: To be used on an experimental basis only in the U.S.

ADDITIONAL INFORMATION: Persists in the soil for 6-12 weeks. Being tested as a seed treatment. Systemic activity.

NAMES

DIOXACARB, C-8353, ELOCRON, FAMID

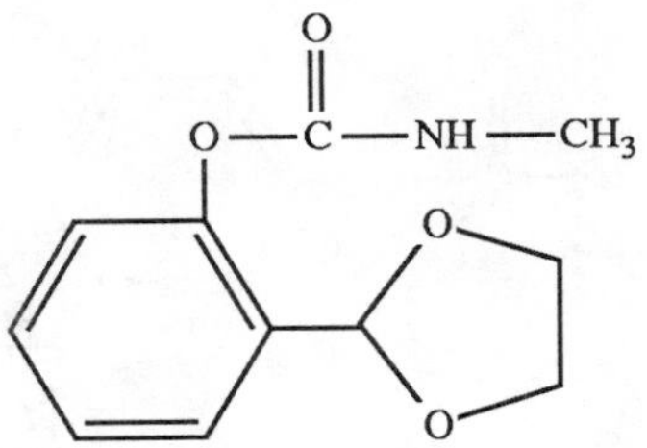

2-(1,3-Dioxolan-2yl) phenylmethyl-carbamate

TYPE: Dioxacarb is a carbamate insecticide, killing as a contact and stomach-poison.

ORIGIN: CIBA-Geigy of Switzerland, 1967.

TOXICITY: LD_{50}-60 mg/kg.

FORMULATIONS: 50% WP, 5% powder, 40% EC.

PHYTOTOXICITY: Appears to be safe on most plants.

USES: Developed for the control of cockroaches and other household pests and foliage pests outside of the U.S. Used on rice, potatoes, rape, and cocoa.

IMPORTANT PESTS CONTROLLED: Cockroaches, Colorado potato beetle, leafhoppers, cocoa bugs, mosquitoes, houseflies, and others.

RATES: Applied 250-750 g a.i./ha.

APPLICATION: Apply when insects appear, and repeat as necessary.

PRECAUTIONS: Not for use in the U.S. Toxic to bees.

ADDITIONAL INFORMATION: Some activity against mites. Compatible with most other pesticides that are not alkaline. Effective against both sucking and chewing insects. Very rapid knockdown effect. Higher rates are required against chewing insects than against sucking ones. Persists on wall surfaces about six months.

NAMES

PROMECARB, CARBAMULT, MINACIDE, PROMECARBE

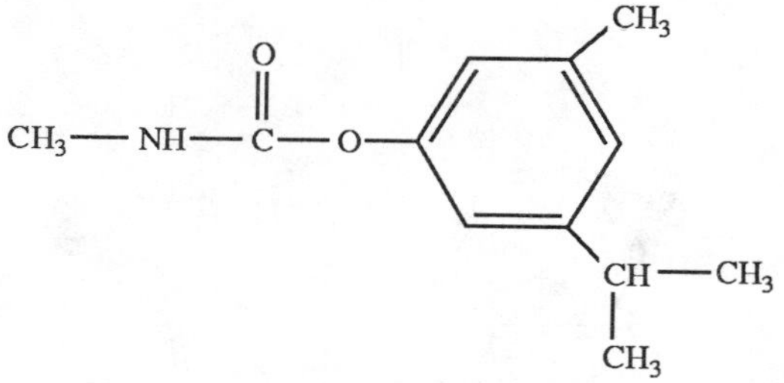

$$CH_3-NH-\overset{\overset{\displaystyle O}{\|}}{C}-O-C_6H_3(CH_3)(CH(CH_3)_2)$$

3-Methyl-5-isopropylphenylmethylcarbamate

TYPE: Carbamult is a carbamate compound used as a contact and stomach-poison insecticide.

ORIGIN: Schering AG of Germany, 1965.

TOXICITY: LD_{50} -74 mg/kg.

FORMULATIONS: 50% WP, 37.5% WP, 5% DUST, 25% EC.

PHYTOTOXICITY: Non-phytotoxic.

USES: Outside the U.S. on potatoes and fruit crops.

IMPORTANT PESTS CONTROLLED: Leaf miners, aphids, flies, mosquitoes, ticks, sawflies, Colorado potato beetles, and many others.

RATES: Applied .1-.25% concentration (50% WP) in water.

APPLICATION: Apply when insects appear and repeat as necessary.

PRECAUTIONS: Do not mix with alkaline materials. Toxic to bees. Not for sale or use in the U.S.

ADDITIONAL INFORMATION: Compatible with most pesticides. Good contact and initial knockdown of insects. Nonsystemic in activity. Primarily controls sucking and chewing insects. Medium residual activity.

NAMES

FENOBUCARB, BPMC, BASSA, BAYCARB, CARVIL, HOPCIN, OSBAC

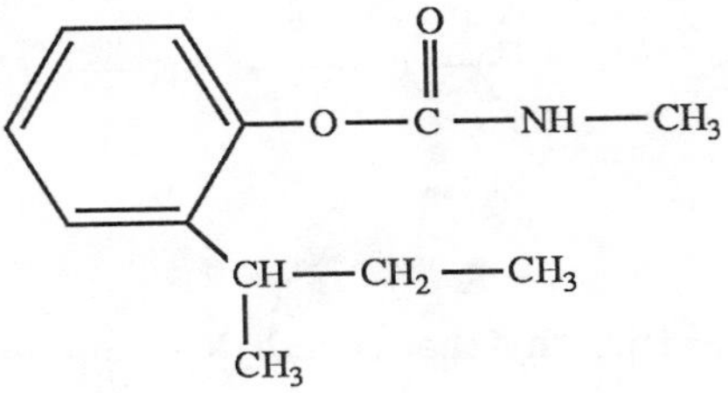

O
‖
—O—C—NH—CH_3

CH—CH_2—CH_3
|
CH_3

2-sec-Butylphenylmethylcarbamate

TYPE: BASSA is a carbamate compound used as a contact insecticide.

ORIGIN: Bayer AG of Germany, Kumiai Chemical Industry Co., Ltd., of Japan. Mitsubishi Chemical Industries and Sumitomo Chemical Co. are the principle basic manufacturers.

TOXICITY: LD_{50}-623 mg/kg.

FORMULATIONS: 50% EC, 3% granules, 2% dust, 500 g/l EC.

PHYTOTOXICITY: Non-phytotoxic to rice.

USES: Outside the U.S. on rice.

IMPORTANT PESTS CONTROLLED: Leafhoppers, aphids, and plant-hoppers.

APPLICATION: Apply to the water and/or plant surfaces whenever insects appear, and repeat as necessary.

PRECAUTIONS: Fish are killed at 24-49 ppm concentration. Not for sale in the U.S.

ADDITIONAL INFORMATION: Kills upon contact with good residual activity. Has a wider application range than most carbamate insecticides used on rice. Mix with BHC to control the rice stem borers.

RELATED MIXTURES:

1. BAYBASSA—A combination insecticide, containing BASSA and BAYCID, developed by Bayer for use on rice in Japan.

NAMES

MPMC, MEOBAL

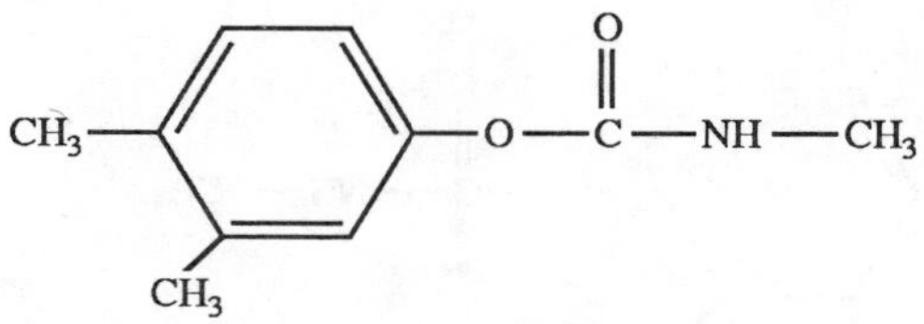

3,4-Dimethylphenylmethylcarbamate

TYPE: MEOBAL is a carbamate compound being used as a foliar insecticide.

ORIGIN: Sumitomo Chemical Co. of Japan, 1967.

TOXICITY: LD_{50}-380 mg/kg.

FORMULATIONS: 50% WP, 2% dust, 3% granule, 30% EC.

PHYTOTOXICITY: Non-phytotoxic when used as directed.

USES: Used in Japan on rice.

IMPORTANT PESTS CONTROLLED: Leafhoppers, scales, and other insects.

RATES: Applied at .05-.1% concentration.

APPLICATION: Apply when insects appear, and repeat as necessary.

PRECAUTIONS: Do not mix with alkaline pesticides.

ADDITIONAL INFORMATION: Moderately toxic to fish. First used commercially in 1967.

NAMES

CARTAP, CADAN, CALDAN, PADAN, PATAP, SANVEX, THIOBEL, VEGETOX

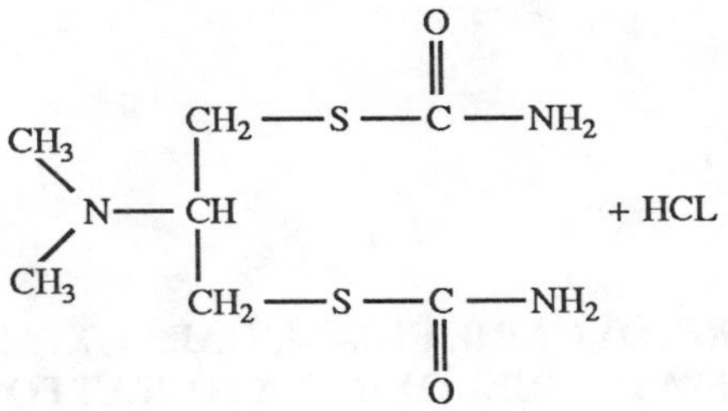

```
                    O
                    ||
          CH2 —S —C —NH2
CH3        |
   \       |
    N— CH                    + HCL
   /       |
CH3        |
          CH2 —S —C —NH2
                    ||
                    O
```

S,S'-(2-(Dimethylamino) trimethylene) bis (thiocarbamate) hydrochloride

TYPE: Padan is an insecticide derived from certain segmented worms, killing upon contact and as a stomach poison.

ORIGIN: Takeda Chemical Industries of Japan, 1965.

TOXICITY: LD_{50}-325 mg/kg.

FORMULATIONS: 25% and 50% water soluble powder, 4% and 10% granules, baits, 2% dusts.

PHYTOTOXICITY: Injury has been noted on cotton and red delicious apples.

IMPORTANT PESTS CONTROLLED: Rice stem borer, leaf rollers, corn borer, boll weevil, Mexican bean beetle, leaf miners, moths, alfalfa weevil, rice borers, armyworms, rice leaf beetle, Colorado potato beetle, flea beetles, codling moth, leafhoppers, aphids, thrips, sawfly, and others.

USES: Used in Japan and other countries on rice, citrus, cotton, potatoes, beans, grapes, tree fruits, apples, vegetables, tea, corn, and other crops.

RATES: Applied at 1/2-2 lb actual/A.

APPLICATION: Apply when insects appear, and repeat as necessary.

PRECAUTIONS: Not for sale or use in the U.S. Toxic to fish. Do not mix with alkaline pesticides.

ADDITIONAL INFORMATION: A derivative of nereistoxin which is a toxin isolated from a marine annelid. Some penetration into the plant tissue can be expected. Compatible with other fungicides and insecticides. Low toxicity to bees. Effective against larvae, adult, and eggs of certain insects. Slow acting. Insects discontinue feeding upon contact with the material, resulting in starvation.

NAMES

BENDIOCARB, **DYCARB, FICAM, GARVOX, MULTAMAT, NIOMIL, ROTATE, SEEDOXIN, TATO, TATTOO, TURCAM**

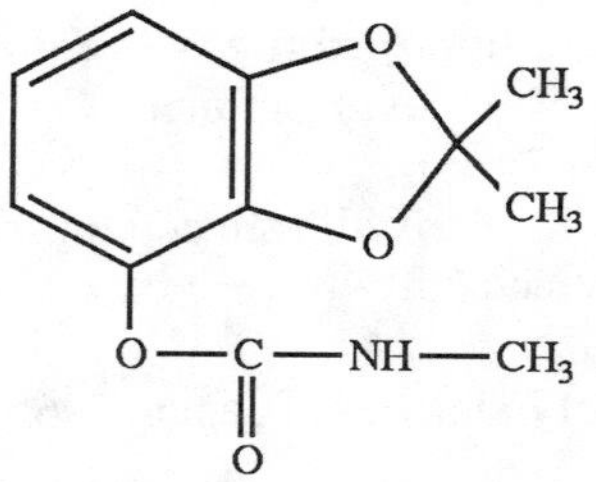

2,2-Dimethyl-1,3-benzodioxol-4-yl methylcarbamate

TYPE: Ficam is a carbamate compound used as a broad-spectrum contact and stomach-poison insecticide.

ORIGIN: Schering Ag of Germany, 1971.

TOXICITY: LD_{50}-40 mg/kg.

FORMULATIONS: 25% ULV, 76% WP, 20% WP, 50% flowable, 3-10% granules.

PHYTOTOXICITY: Non-phytotoxic when used as directed. Do not use on coleus.

USES: Used for the control of household pests and in food handing establishments, premises, non-bearing fruit trees, and ornamentals. Being tested as a soil insecticide. Used outside the U.S. as a seed treatment on sugar beets, as well as for many other purposes.

IMPORTANT INSECTS CONTROLLED: Ants, bed bugs, carpet beetles, cockroaches, corn rootworms, crickets, earwigs, fleas, mosquitoes, silverfish, spiders, termites, wasps, wireworms, and agricultural and veterinary pests.

APPLICATION: Apply when insects appear, and repeat as necessary. Apply to the soil preplant. Used as a seed treatment. Used as a ULV spray for mosquito control.

RATES: Applied at .25-.5% ai. Apply to the soil at .75-1 lb a.i./A.

PRECAUTIONS: When added to water, the spray solution changes color from white to brown after a few days. Discard the solution when this happens. Highly toxic to fish.

ADDITIONAL INFORMATION: Rapid knockdown with good residual activity. 10 weeks' control can be expected. Has no odor and does not stain the treated surface. Gives 6-8 weeks' control as a seed treatment. Limited systemic activity.

NAMES

***BENFURACARB,* ONCOL, OK-174**

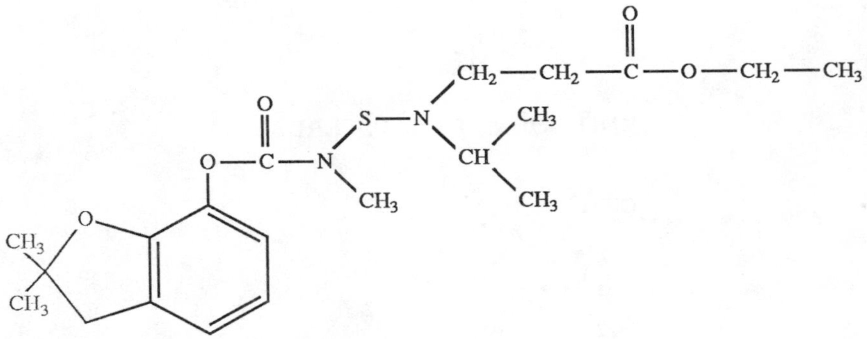

2,3-Dihydro-2,2-dimethyl-7-benzo furanyl N-[N-[2-(ethylcarbonyl) ethyl]-N-isopropyl sulfenamoyl]-N-methylcarbamate

TYPE: Oncol is a carbamate compound used as stomach-poison and contact insecticide.

ORIGIN: Otsuka Chemical Co. of Japan, 1981.

TOXICITY: LD_{50}-138 mg/kg. May cause some skin irritation.

FORMULATION: 3, 5, and 10% granules, 20 and 30% EC, 60% liquid.

PHYTOTOXICITY: Non-phytotoxic when used as directed.

USES: Experimentally being tested on corn, rice, sorghum, sugar beets, potatoes, soybeans, fruit trees, citrus, cotton, and others.

IMPORTANT INSECTS CONTROLLED: Wireworms, leafhoppers, aphids, corn borers, armyworms, rice water weevil, flea beetles, Colorado potato beetle, cutworms, cabbageworm, codling moth, scale, oriental fruit moth, thrips, nematodes, and others.

RATE: Applied at .4-2 kg a.i./ha.

APPLICATION: Apply as a soil treatment or as a foliar spray depending upon the insects being controlled.

PRECAUTIONS: Not for sale or use in the U.S.

ADDITIONAL INFORMATION: Fish are moderately tolerant of the material. Noncorrosive.

NAMES

KNOCKBAL, TBPMC, TERBAM

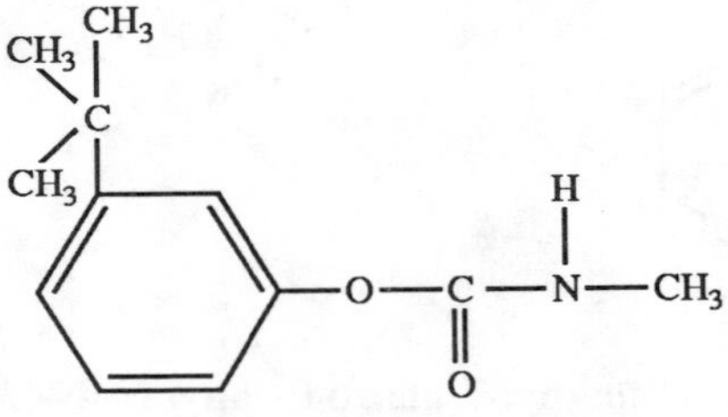

CH_3 CH_3 C CH_3 H O—C—N—CH_3 O

3-tert. -Butylphenyl-N-methylcarbamate

TYPE: Knockbal is a carbamate compound used as a contact insecticide.

ORIGIN: 1970-Hokko Chemical Co. and Hodogaya Chem. Co. of Japan.

TOXICITY: LD_{50}-470 mg/kg. (mice).

FORMULATIONS: 2% dust, 50% WP.

PHYTOTOXICITY: Non phytotoxic when used as directed.

IMPORTANT PESTS CONTROLLED: Planthoppers, leafhoppers, scales, tortrix mites and many others.

USES: In Japan on rice, peaches, tea, and forest trees.

PRECAUTIONS: Toxic to fish. Not for sale or use in the U.S.

ADDITIONAL INFORMATION: A wide spectrum insecticide.

NAMES

MEXACARBATE, ZECTRAN

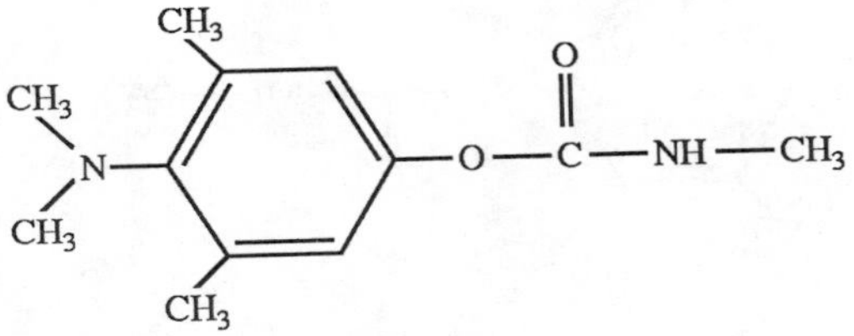

CH_3 O CH_3 N— —O—C—NH—CH_3 CH_3 CH_3

4-(Dimethylamino)-3,5-xylyl methylcarbamate

TYPE: Zectran is a carbamate compound used as a contact, systemic and stomach-poison insecticide-acaricide.

ORIGIN: Dow Chemical Co., 1961. Now being developed by Rhone Poulenc.

TOXICITY: LD_{50}-15 mg/kg.

FORMULATION: 2 EC.

PHYTOTOXICITY: Injury has been reported on certain ornamentals such as maidenhair fern, cape chestnuts, Geraldron wax flowers, and certain varieties of roses, phlox, and petunias.

USES: Used on ornamental flowers, shrubs, vines, trees, evergreens, turf, and ground covers.

IMPORTANT PESTS CONTROLLED: Loopers, caterpillars, armyworms, cutworms, leafminers, thrips, lygus, leafhoppers, aphids, scales, mealybugs, whiteflies, mites, snails, slugs, sod webworms, millipeds, sowbugs, and others.

RATES: Applied at .5-.66 lb a.i./100 gal of water.

APPLICATION: Apply when insects appear and repeat as necessary.

PRECAUTIONS: Do not use on any food or feed crops. Toxic to wildlife.

ADDITIONAL INFORMATION: Compatible with other insecticides. Snails and slugs will stop feeding soon after treatment but complete kill may take 2-3 days. May be used on both greenhouse and outdoor ornamentals.

NAMES

MACBAL, XMC

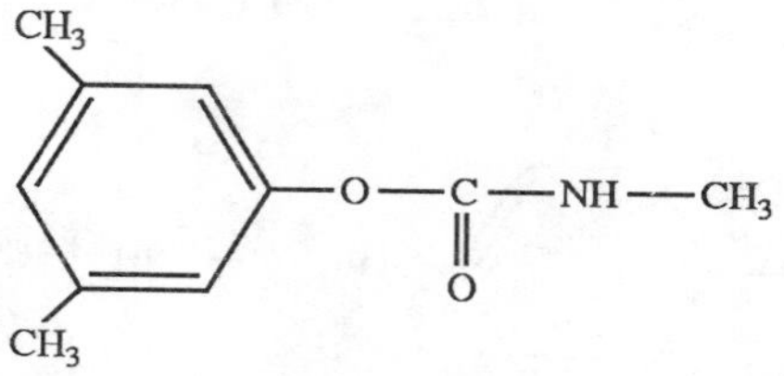

CH_3 CH_3 O—C(=O)—NH—CH_3

3,5-Xylyl-N-methylcarbamate

TYPE: Macbal is a carbamate insecticide applied to the foliage.

ORIGIN: Hodogaya Chemical Co. of Japan, 1969.

TOXICITY: LD_{50}-542 mg/kg.

FORMULATIONS: 20% EC, 50% WP, 2% dust.

PHYTOTOXICITY: Non-phytotoxic when used as directed.

USES: Used on rice in Japan.

IMPORTANT PESTS CONTROLLED: Leafhoppers, and planthoppers, snails, and slugs.

APPLICATION: Apply when insects appear, and repeat as necessary.

PRECAUTIONS: Not for sale in the U.S. Avoid spraying closely before or after a propanil application.

ADDITIONAL INFORMATION: Relatively Non-toxic to fish. Has a fast knock-down effect.

NAMES

FENOXYCARB, **INSEGAR, LOGIC, PICTYL, RO-13-5223, TORUS, VARIKILL**

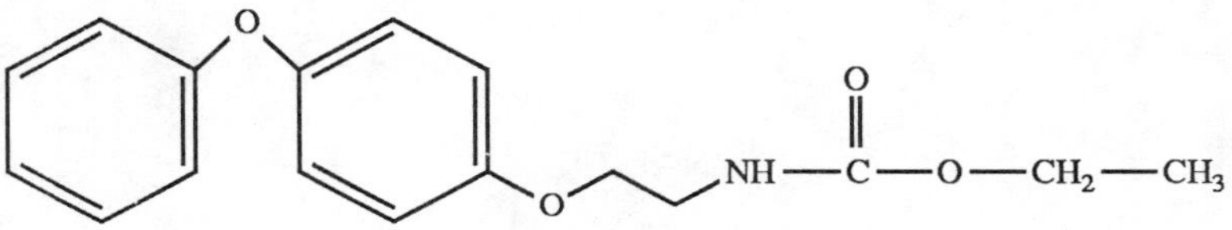

Ethyl [2-phenoxyphenoxy)ethyl]carbamate

TYPE: Fenoxycarb is a carbamate compound used as an insect growth regulator either by contact or as a stomach-poison.

ORIGIN: Dr. R. Maag Ltd. of Switzerland, 1982. Being developed in the U.S. by Maag Agrochemicals.

TOXICITY: LD_{50}-9220 mg/kg.

FORMULATION: 25% WP, 1% bait, 5% granules, 5% dust 125 EC, 2 EC.

PHYTOTOXICITY: Some varieties of pears, grapes, citrus, and ornamentals have been injured.

USES: Non crop areas, turf and nonbearing citrus. Experimentally being used in forestry, stored products, apples, grapes, citrus, cotton, pears, ornamentals, as a public health insecticide, and others.

IMPORTANT INSECTS CONTROLLED: Ants, roaches, fireants, fleas, ticks, chiggers, stored product insects, termites, spruce budworm, mosquitoes, gypsy moth, bollworms, cotton leaf perforator, pear psylla, scales, leafminers, and others.

RATES: Various, depending on the application and insect.

APPLICATION: Used as a bait, a crack and crevice dust, a stored product spray application and as a foliar spray.

PRECAUTION: Within 2 hours of treatment, the foraging ants gateher the bulk of the bait and bring it back to the mound where they feed on it. Toxic to fish.

ADDITIONAL INFORMATION: Exhibits strong juvenile hormone activity which induces ovicidal effects, inhibits metamorphosis of the adult stage (death in the last larvae or pupae stage) and interferes with the molting of early instar larvae. Quite specific in activity. Not active on honeybees. Rapid dissipation in the soil or on plants. No immediate killing effects are seen. Multiple treatments may be necessary. Induces the queen to produce only non-worker ants that are incapable of developing into adults. Slow acting requiring 2-4 weeks of control.

NAMES

HEXYTHIAZOX, CESAR, NA-73, NISSORUN

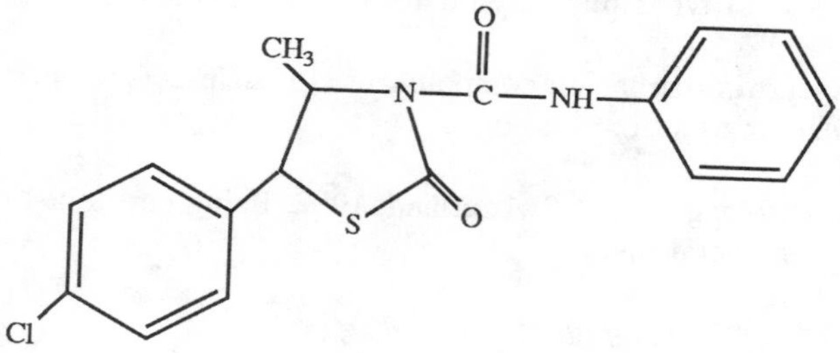

CH3 O N C NH S O Cl

trans-5-(4-Chlorophenyl)-N-cyclohexyl-4-methyl-2 oxothiazolidinone-3-carboxamide

TYPE: Nissorun is an organic compound used as a contact and stomach-poison acaricide.

ORIGIN: Nippon Soda of Japan, 1980.

TOXICITY: LD_{50}-5000 mg/kg.

FORMULATIONS: 10 and 50% WP, 10% EC.

PHYTOTOXICITY: Non-phytotoxic when used as directed.

IMPORTANT INSECTS CONTROLLED: Mites

USES: Used outside the U.S. on apples, citrus, corn, grapes, vegetables, cotton, and other crops.

RATES: Applied at 2.5-5 g a.i./ha or at a concentration of 30-50 ppm.

APPLICATION: Apply early in the season before mites build up. Repeat as necessary.

PRECAUTIONS: Not for sale or use in the U.S. Not effective on adult mites.

ADDITIONAL INFORMATION: Nonsystemic. Compatible with other pesticides. Good ovicidal activity and the eggs laid by treated adult females will not be fertile. Non-toxic to bees.Effective on all stages of mite growth. Control lasts for 50-60 days.

NAMES

BUTOCARBOXIME, DARWIN 755

CH_3

CH_3—S—CH—C—CH_3

N—O—C—NH—CH_3

O

3-Methylthio-o-[(methylamino)carbomyl]oxime-2 butanone

TYPE: Butocarboxime is a carbamate compound used as a contact and stomach-poison insecticide.

ORIGIN: Wacker-Chemie GmbH of West Germany, 1973.

TOXICITY: LD_{50}-158 mg/kg.

FORMULATIONS: 500 g a.i./l EC, 50g/1 EC, and .8g/1 EC.

USES: Outside the U.S. on fruit crops, vegetable crops, ornamentals, field crops, tobacco, cotton, hops, and others.

IMPORTANT PESTS CONTROLLED: Aphids, scales, mealy bugs, mites, white flies, thrips, and others.

RATES: Applied at 1/4-1 lb a.i./100 gal of water or 2-1/4 -3- 3/4 lb a.i./A.

APPLICATION: Apply when insects appear, and repeat as necessary.

PRECAUTIONS: Not for sale or use in the U.S. Toxic to bees.

ADDITIONAL INFORMATION: High mite populations are only suppressed, not controlled. Harmless to predaceous and parasitic anthropods. Persists in the plant for 15-20 days.

NAMES

BUTOXYCARBOXIM, BUG PIN, PLANT PIN

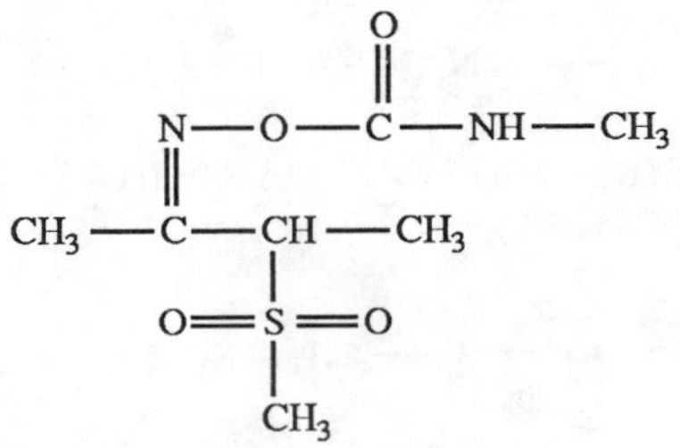

```
                O
                ‖
        N—O—C—NH—CH3
        ‖
CH3—C—CH—CH3
             |
        O═S═O
             |
            CH3
```

3-(Methylsulfonyl)-O-[(methylamino)carbonyl]oxine-2-butanone

TYPE: Butoxicarboxim is a carbamate compound used as a contact and stomach-poison insecticide.

ORIGIN: Wacker-Chemie GmbH of West Germany, 1972. Sold in the U.S. by Knoll Bioproducts.

TOXICITY: LD_{50}-458 mg/kg.

FORMULATIONS: Incorporated into pasteboardpins containing 50 mg ai.

USES: Outside the U.S. on ornamentals.

IMPORTANT PESTS CONTROLLED: Aphids, mites, thrips, and other sap-sucking insects.

APPLICATION: Insert into the soil close to the stem of the potted plants. Because of the water solubility, the active ingredient is rapidly released in the moisture of the soil and absorbed by the plants' root systems.

PRECAUTIONS: Not for sale or use in the U.S.

ADDITIONAL INFORMATION: A translocated, root-absorbed insecticide. The insecticide effects can be observed in 3-7 days, and reach the full extent in 7-14 days. Control should last for 4-8 weeks, until the cardboard strips rot. Used both for home-potted plants and for the commercial-greenhouse grower.

NAME

WL-108477

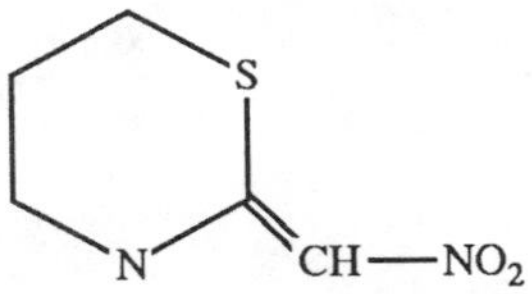

S

N CH—NO_2

2-Nitromethylene-1,3-thiazinan-3,-yl-carbamaldehyde

TYPE: WL-108477 is a nitromethylene heterocyclic compound used as a fast acting insecticide.

ORIGIN: Shell Chemical Co. of England, 1982.

TOXICITY: LD_{50} 1000 mg/kg (mouse).

PHYTOTOXICITY: Non-phytotoxic when used as directed.

FORMULATION: 25% WP.

USES: Experimentally being tested on rice, vegetables and other crops.

IMPORTANT INSECTS CONTROLLED: Plant hoppers, cotton leafworm, leaf-hopper, stink bugs, velvet bean caterpillar, white fly and others.

RATES: Applied at 100-300 g. a.i./ha.

APPLICATION: Apply when insects appear and repeat as necessary.

PRECAUTION: Use on an experimental basis only.

ADDITIONAL INFORMATION: A new class of insecticides. Extremely rapid activity. Is not rapidly broken down by sunlight. Low fish toxicity. Short persistence.

ANIMAL PLANT DERIVATIVES
SYNTHETIC PYRETHROIDS
and INORGANIC COMPOUNDS

NAMES

CAMPHECHLOR, TOXAPHENE, ATTAC, MOTOX, PHENACIDE, PHENATOX, POLYCHLORO-CAMPHENE, STROBANE-T, TOXAKIL

$C_{10}H_{10}C_{18}$

(Technical grade of chlorinated camphene containing 67-69% chlorine.) Octachlorocamphene

TYPE: Toxaphene is a chlorinated-hydrocarbon insecticide derived from the Southern pine, showing contact and stomach-poison activity.

ORIGIN: Hercules Inc. (now Nor-Am Chemical Co.), 1946. No longer sold in the U.S.

TOXICITY: LD_{50}-49 mg/kg. Readily absorbed through the skin.

FORMULATIONS: WP 40%, 4, 6, and 8 EC, 10% and 20% granules, 1% baits. Also mixed with other pesticides.

PHYTOTOXICITY: Injury has been reported on cucurbits and Imperial Gage plums. It may cause off-flavor in stored tobacco.

USES: Cotton, corn, small grains, peanuts, soybeans, sheep, and cattle.

IMPORTANT PESTS CONTROLLED: Leafhoppers, loopers, armyworms, corn earworms, cutworms, grasshoppers, European cornborer, lygus bugs, aphids, spittlebugs, thrips, leaf miners, chinch bugs, boll weevils, bollworms, chiggers, crickets, ticks, and many others.

RATES: Applied at 1-8 lb actual/100 gal water.

APPLICATION:

1. Foliage-Apply at a uniform rate with common application equipment. Repeat as necessary.
2. Soil-Band treatment mixed with at least 10 gal water/A, applied in the late evening or at night. Cultivate (throwing soil over the row) immediately after treatment.
3. Livestock-Treat by spraying, dipping, or with backrubbers. Do not treat within 28 days of slaughter.

PRECAUTIONS: Corrosive to iron. Lactating dairy animals should not be treated. Toxic to fish. Noncompatible with Bordeaux, calcium arsenate, ferric sulfate, lime, and lime sulfur. Birds feeding on treated areas may be killed. No longer used in the U.S.

ADDITIONAL INFORMATION: Broken down by strong alkalies. Accumulation in the body fat dissipates quickly. Decomposed by soil organisms. Used to control certain weeds in soybeans in some areas.

NAMES

BACILLUS THURINGIENSIS-BERLINER, **AGRITOL, AATACK, BACTIR, BACTOSPEINE, BIOBIT, BIOTROL, BTB, BTV, BUG-TIME, CONDOR, DELFIN, DIPEL, FORAY, JAVELIN, LARVATROL, LARVO-BT, LASER, LEPTOX, NOVABAC-3, SOK-BT, SPOREINE, STAN-GUARD, THURICIDE, TRIBACTUR**

Bacillus thuringiensis var. kurstaki

TYPE: Bacillus thuringiensis is a bacterial organism which causes disease in certain insects, thereby controlling them.

ORIGIN: Nutrilite Products, Inc., 1961. Sondoz, Novo, Abbott Labs, PBI Gordon and others sell this material in the U.S.

TOXICITY: Non-toxic to animals

FORMULATIONS: WP, flowable formulations, and aqueous concentrates.

PHYTOTOXICITY: Non-phytotoxic.

USES: Alfalfa, almonds, apples, artichokes, bananas, beans, beets, broccoli, Brussels sprouts, cabbage, cauliflower, celery, coffee, collards, corn, cotton, cucumbers, eggplant, garlic, grapes, horseradish, kale, lettuce, melons, mint, mustard, nectarines, onions, oranges, parsley, peaches, peanuts, pears, peas, peppers, potatoes, pumpkins, radishes, soybeans, spinach, squash, strawberries, sugarbeets, Swiss chard, tobacco, tomatoes, turf, turnips, walnuts, watermelons, and ornamentals. Controls the wax moth larvae in honeycombs

IMPORTANT PESTS CONTROLLED: Cabbage loopers, Spruce budworm, imported cabbageworms, gypsy moth, tobacco hornworms, artichoke plume moths, armyworms, and the larval stages of other lepidoptera insects with a high gut pH.

RATES: Applied at 1/2-4 lb actual/A in sufficient water for thorough coverage.

APPLICATION: Apply when insects first appear and repeat at weekly intervals or as often as necessary. Suggested for use as a replacement or a supplement to chemical insecticides in situations where residues may be undesirable. May be applied at any time during the growing period. Agitate while spraying. Apply in late evenings in hot, dry areas. May be applied through sprinkler irrigation systems.

PRECAUTIONS: Do not expect the insect to die immediately. They remain on the plant, but stop eating, until they starve to death. Do not allow dilute sprays to stand in the spray tank for more than 12 hours. Do not store at above 90°F.

ADDITIONAL INFORMATION: Compatible with other pesticides. It must be eaten by insects to become fatal to them. It affects the insects by paralyzing their stomachs. The insects then remain on the plant for 24-48 hours until starvation causes death. Remains effective for several days following application. It may need a spreader-sticker added. The only insects controlled are caterpillars having an alkaline gut. Best results are obtained when applied early. Used in forests for insect control. May be applied by air. Non-toxic to predators of the susceptible lepidoptera larvae. The Javelin formulation of B.T. by Sandoz controls beet armyworm.

NAME

CERTAN

Bacillus Thuringiensis Berliner

TYPE: Certan is a bacterial organism that is used as a biological insecticide.

ORIGIN: Sandoz Crop Protection 1980.

TOXICITY: Non-toxic.

FORMULATIONS: .8% active liquid.

USES: To control the wax moth larvae in honeycombs.

APPLICATIONS: Apply in the fall prior to placing dried combs in storage or in the spring before returning frames to the colony. Apply from the wax moth egg hatch to the second instar larvae. Apply directly to each side of the comb.

PRECAUTIONS: Thorough coverage is essential. Do not store above 80° F.

ADDITIONAL INFORMATION: Must be eaten by the larvae to be effective. Causes no detrimental effect on the bee colony or the honey produced.

NAMES

BACTIMOS, BMC, BTI, TEKNAR, VECTOBAC

Bacillus thuringiensis Berliner var. israelensis

TYPE: BTI is a biological compound used as a larvicide for mosquito control.

ORIGIN: Sandoz, Abbott, and Novo BioKentrol, 1980.

TOXICITY: Non-toxic to warm-blooded animals.

FORMULATIONS: .8% active (1500 AA units/mg) or 50% WP (3500 AA units/ml), granules, flowables, briquets.

USES: Ponds, ditches, tidal water, catch basins, sewage areas, and other areas where water stands.

IMPORTANT INSECTS CONTROLLED: Mosquitoes, fungus gnat and black fly.

RATES: Applied at 1/2-2 pt/A, or 2-8 oz/A of the 50% WP.

APPLICATION: Apply by ground or air to the water surface. Use the lower rates when the 1st to 3rd instar larvae are present and the higher rate for the 3rd to 4th instar. Re-apply as needed.

PRECAUTIONS: Do not let stand in the spray tank for over 12 hours. Do not store at temperatures over 90°F. Brackish waters may take a higher rate.

ADDITIONAL INFORMATION: The microorganisms released into the environment do not self-perpetuate, so re-application is required. Harmless to people, animals, fish, plants, and other non-target insects. After ingestion by the larvae, it causes paralysis of the midgut to take place within a few hours, and death will occur within 24 hours. Residual activity is rarely longer than 24 hours. Not effective on the pupal or adult stages.

NAMES

BT STRAIN EG 2348, CONDOR

Bacillus thuringiensis strain EG 2348

TYPE: Condor is a biological compound used as a larvicide.

ORIGIN: Ecogen, Inc. 1986.

TOXICITY: Non-Toxic to animals.

FORMULATIONS: .64 lbs. a.i./gal. oil flowable.

PHYTOTOXICITY: Non-phytotoxic.

IMPORTANT INSECTS CONTROLLED: Gypsy moth, spruce budworm.

USES: Forestry.

RATES: Applied at 1-3 pints/acre.

APPLICATIONS: Applied by ground or air when insects appear. For spruce bud worm, apply when 50% of the larvae are in the 3rd or 4th instar.

PRECAUTIONS: Slow acting. Do not mix more product with water than can be used in a 144 hour period.

ADDITIONAL INFORMATION: A spreader/sticker may be added. Thorough coverage is necessary. Most effective on young larvae before extensive foliar damage occurs.

NAMES

B.T.T., TRIDENT

Bacillus thuringiensis var. tenebrionis

TYPE: B.T. var. tenebrionis is a biological compound used as a larvacide for the Colorado potato beetle.

ORIGIN: Sandoz Crop Protection 1986.

TOXICITY: Non-toxic to animals.

FORMULATION: 3200 units/milligram liquid.

PHYTOTOXICITY: Non-phytotoxic.

IMPORTANT INSECTS CONTROLLED: Colorada potato beetle.

USES: Potatoes.

RATES: Apply 4-6 quts./acre in 20 gallons of water by ground or 3-10 gallons by air.

APPLICATIONS: Apply to the plant foliage. When the larvae ingests the product they stop eating. Apply when the larvae are 1/4 inch in length. Repeat as necessary.

PRECAUTIONS: Slow acting Do not allow diluted sprays to remain in the tank over 72 hours.

ADDITIONAL INFORMATION: Most effective on young larvae. May be mixed with other pesticides. A spreader may be added to improve control. Larvae stop feeding within a few hours and die within 2-5 days. Thorough coverage of the foliage is essential.

NAMES

***B.T.s.d.*, M-ONE**

Bacillus thuringiensis var. san diego

TYPE: M-One is a biological compound used as a larvacide to control the Colorado potato beetle.

ORIGIN: Mycogen Corp. 1986.

TOXICITY: Non-toxic to animals.

FORMULATION: 4.5% water dispensible liquid.

PHYTOTOXICITY: Non-phytotoxic.

IMPORTANT INSECTS CONTROLLED: Colorado potato beetle.

USES: Potatoes, tomatoes and eggplant.

RATES: Applied at 2-6 qts/acre.

APPLICATION: Apply when egg hatch begins. Continue spraying on a 3-10 day basis depending upon the population. May be applied by air.

PRECAUTIONS: Use within 12 hours of mixing. Re-apply if rain occurs within 24 hours. Store at temperatures of 35° F to 90° F.

ADDITIONAL INFORMATION: Dissolves in the insect digestive tract causing it to immediately stop feeding. Death occurs within 2-5 days. A very specific insecticide. The adult Colorado potatoe beetle is not controlled.

NAMES

B.T. strain EG 2424, FOIL

Bacillus thuringiensis strain EG 2424

TYPE: B.T. strain EG 2424 is a bacterial organism used as a biological insecticide.

ORIGIN: Ecogen, Inc. 1986.

TOXICITY: Non-toxic to warm blooded animals.

FORMULATION: 7.5% active liquid.

USES: Experimentally on potatoes to control the Colorado potatoe beetle and the European corn borer.

RATES: Applied at 1-3 qts. formulation/acre.

APPLICATION: Apply by ground or air to give thorough coverage. Treat when the larvae are young. May be applied with a spreader/sticker. Repeat application as necessary.

PRECAUTIONS: Used on a experimental basis only. Do not mix more product than can be used in a 144 hour period.

ADDITIONAL INFORMATION: Slow acting. Insects stop eating once they ingest this material.

NAMES

THURINGIENSIN, DIBETA

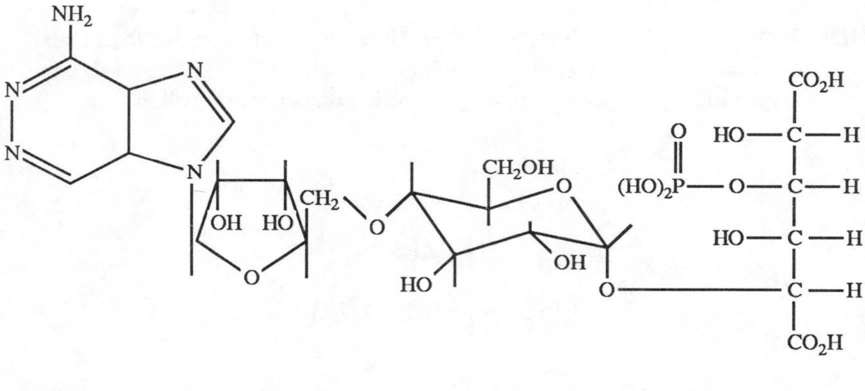

TYPE: DiBeta is a biological insecticide derived from the bacterial fermentation of a strain of Bacillus thuringiensis.

ORIGIN: Abbott Labs 1984.

TOXICITY: LD_{50}-18,700 mg./kg. May cause skin and eye irritation.

FORMULATIONS: 1.5% liquid.

PHYTOTOXICITY: Some injury has been reported on certain potato varieties and on chrysanthemums.

IMPORTANT INSECTS CONTROLLED: Mites, lygus bugs, Colorado potatoe beetle, house flies, plant bugs, armyworms and others.

USES: Experimentally being tested on cotton, potatoes, ornamentals and other crops.

RATES: Apply at 2-4 qts./acre.

APPLICATION: Apply when insect population begins to build and repeat as necessary.

PRECAUTIONS: Used on an experimental basis only. Slightly toxic to fish. Not compatable with spray oils.

ADDITIONAL INFORMATION: Also known as beta-exotoxin. Controls both the nymphs and adult mites and the larvae of Colorado potato beetle. Compatible with other pesticides. Used primarily as a stomach poison with contact activity on mites. Slow acting.

NAMES

DOOM, JAPIDEMIC, MILKY SPORE

Contains spores of Bacillus popilliae Dutky

TYPE: Milky Spore is a biological product which contains live disease spores.

ORIGIN: Fairfax Biological Labs and Reuter Labs.

TOXICITY: Non-toxic to warm blooded animals.

FORMULATION: Numerous-Not less than 100 million viable spores/gram of powder, granules.

PHYTOTOXICITY: Non-phytotoxic.

USES: Turf and grassy areas.

IMPORTANT INSECTS CONTROLLED: Japanese beetle larvae.

RATES: Use at 10 lbs/A. (.016 ai)

APPLICATION: Apply to turf or grassy areas and water for 20-30 minutes to allow the spores to enter the soil.

PRECAUTION: Specific for control of the grub stage only.

ADDITIONAL INFORMATION: The disease these spore cause is named as such because of the increased whiteness of the grub as the bacteria multiply within its body. The infested grubs die releasing additional spore into the soil multiplying the effects. Remain persistent in the soil even under adverse environmental conditions. Allow 3-4 years for the optimum spreading of the spores throughout the soil. A self perpetuating microbial insecticide.

NAMES

GRASSHOPPER SPORE, HOPPER STOPPER, NOLOBAIT, NOLOC, SEMAPORE, TROJAN-10

1 X 10^9 Nosema locustae spores per ml.

TYPE: Noloc is a biological insecticide consisting of a natural disease that attacks grasshoppers.

ORIGIN: Reuter Labs, Evans Biocontrol and others. 1979.

TOXICITY: No effects on warm-blooded animals.

PHYTOTOXICITY: Non-phytotoxic.

FORMULATIONS: 7.5% concentrate, .0459% powder.

IMPORTANT INSECTS CONTROLLED: Grasshoppers, and some crickets.

USES: Rangeland and noncrop areas. Also for home garden usage and all agricultural commodities.

APPLICATION: Must be consumed to be infective. Spray the spores on to wheat bran and distribute the bran by air or ground at 1 lb/A. Once the bran is treated, apply it as soon as possible. Also spread by hand or broadcast equipment for small areas. Treat grasshoppers when they are in the third instar.

PRECAUTIONS: Slow-acting.

ADDITIONAL INFORMATION: Attacks the adipose tissue of grasshoppers, depriving them of the energy to grow and develop. Spores are released into the environment when the infected insect dies. Since cannibalism is characteristic of this species, they continue to spread the disease. The natural occurrence of this disease in grasshoppers is less than 1%. Up to 50% reduction in grasshopper populations has been obtained 3-4 weeks after application. Eggs laid in the fall often carry the disease with them. Not too effective on adults.

NAME

DECYDE

Granulosis virus of codling moth larvae

TYPE: DECYDE is a biological compound used as a stomach poison insecticide.

ORIGIN: Being developed by the University of California.

TOXICITY: Non-toxic to warm blooded animals.

FORMULATION: .05% active powder.

IMPORTANT INSECTS CONTROLLED: Codling moth.

USES: Experimentally on walnuts, pears and apples.

RATES: Applied at .5-2 lb. formulation/acre.

APPLICATION: Apply at the beginning of egg hatch and repeat at 6-10 day intervals as long as oviposition takes place. A spreader/sticker improves the effectiveness.

PRECAUTIONS: Used on an experimental basis only. Do not use chlorinated water in the spray tank. Do not expose to temperatures above 90° F. Do not let stand in the spray tank for over 24 hours.

ADDITIONAL INFORMATION: A naturally occurring disease agent. Store in a freezer.

NAME

VECTOLEX

Bacillus Sphaericus

TYPE: Vectolex is a biological insecticide being developed for use on mosquitos.

ORIGIN: Abbott Labs 1986.

TOXICITY: Non-toxic to warm blooded animals.

FORULATIONS: Granules, WP and aqueous suspension.

USES: Being developed as a mosquito larvicide.

PRECAUTIONS: To be used on an experimental basis only.

ADDITIONAL INFORMATION: Controls mosquitos that develope in a dirty water habitat.

NAMES

VERTALEC, MYCOTAL

spores of the fungus Verticillium lecanii

TYPE: Vertalec/Mycotal is a biological compound used as an insecticide.

ORIGIN: Microbial Resources 1984.

TOXICITY: Non-toxic to warm blooded animals.

FORMULATIONS: Various.

PHYTOTOXICITY: Non-phytotoxicity has been noted when used as directed.

USES: Mycotal — Used to control white flies in glass houses that attack such crops as cucumbers, tomatoes, peppers and ornamentals.

Vertalec — Used in glass houses to control aphids on numerous crops.

RATES: Applied at 2.5 g/l in 500-1000 l/ha.

APPLICATIONS: Use only in glass houses where night temperatures are kept at 15-25° C and the humidity above 85%. Spray on, giving thorough coverage.

PRECAUTIONS: Used on an experimental basis only. Do not mix with fungicides.

ADDITIONAL INFORMATION: Specific in activity. Infected insects die in 10-14 days and act as a reservoir of infection giving long lasting control.

NAMES

BIOSAFE, HP 88, NC 500, NC ALL, HL-88

Naturally occurring entomogenous nematodes

TYPE: Biosafe is a naturally occurring organism used to control insects.

ORIGIN: Biosys, Inc. 1985.

TOXICITY: Non-toxic to warm blooded animals.

FORMULATIONS: Various.

IMPORTANT INSECTS CONTROLLED: Japanese beetle, sod webworms. cutworms, white grubs, wireworms, armyworms, fire ants, weevils, fleabeetles, mole crickets and others.

USES: Vegetables, turf, ornamentals.

APPLICATION: Apply to the soil either preplant or to growing plants. Water the soil after application so it is saturated. Repeat in 3-4 weeks.

PRECAUTIONS: Do not freeze, but store in a refrigerator. Use within 45 minutes after mixing with water. Avoid direct sunlight when applying.

ADDITIONAL INFORMATION: A number of nematode species specific to controlling certain insects are being worked with. Effective on soil dwelling insects. The insect should be killed within 48 hours.

NAME

VIROX

Milled larvae of Neodiprion sertifer containing NVP.

TYPE: Virox is a microbial insecticide used to control European pine sawfly.

ORIGIN: Microbial Resources 1985.

TOXICITY: Non-toxic to warm blooded animals.

FORMULATION: 2% liquid.

PHYTOTOXICITY: Non-phytotoxic when used as directed.

USES: Used on pine trees to control the pine sawfly.

RATES: Apply at 1.4 fl. ounces/acre.

APPLICATION: Apply by ground or air. Most effective against small larvae so try to apply at 90% egg hatch.

PRECAUTIONS: Do not mix with other pesticides. Do not leave in the spray tank longer than 12 hours.

ADDITIONAL INFORMATION: Produced from the larvae of pine sawfly (Neodriprion sertifer) that has been previously infected with NPV virus. Following consumption, it attacks the insects stomach wall reducing the insects feeding and killing it within 10-21 days. Specific in activity.

NAMES

D-LIMONENE, DEMIZE

1,8(9)-p-Methadiene 1-methyl-4-isopropenyl-1-cyclohexene

TYPE: D-Limonene is an organic botanical compound used as a contact insecticide.

ORIGIN: Pet Chemicals, Inc. 1981.

TOXICITY: LD_{50} -5,000 mg/kg or higher. May cause eye irritation.

FORMULATION: An aerosol spray, a pet shampoo, a dip and pre-mixed pump spray.

IMPORTANT INSECTS CONTROLLED: Fleas and ticks.

USES: Apply as a direct application to pets and their bedding.

PRECAUTIONS: Do not use on animals less than 6 weeks old.

ADDITIONAL INFORMATION: Mode of action similar to pyrethrin. Citrus scented. May be mixed with other insecticides.

NAMES

SAFER INSECTICIDAL SOAP

Potassium salt of fatty acids

TYPE: Insecticidal Soap is an organic compound used as a contact insecticide.

ORIGIN: Safer, Inc. 1985.

TOXICITY: Non-toxic.

FORMULATION: 51% active liquid. 1% liquid.

PHYTOTOXICITY: Injury has occured on some ornamentals such as horse chestnut, mountain ash, Japanese maple, gardinia, bleeding heart, sweet pea, maidenhead fern, crown of thorns, some poinsettias and others.

IMPORTANT INSECTS CONTROLLED: Adelgids, aphids, earwigs, grasshoppers, lace bugs, mealy bugs, mites, plant bugs, psyllids, scales, thrips, white flies and others.

USES: Can be used on foliage plants, flowers, shrubs, trees, fruits, nuts and vegetables.

RATES: Applied at 2.5 fl. oz.-4 fl. oz/gallon of water.

APPLICATION: Apply when insects first appear. Wet thoroghly all infested plant surfaces.

PRECAUTINOS: Do not use on Easter lilies during bud formation. Avoid spray when blossoms are present. Most effective in soft water, so a chelating agent (EDTA) may need to be added to hard water. Foams when mixed with water. Do not mix with fertilizers, lime sulfurs, copper sulfate, Bordeaux or Rotenone.

ADDITIONAL INFORMATION: Made from naturally occuring biodegradable fatty acids. Also used as a moss and algae killer on structural surfaces and a moss killer on turf. A natural spreader/sticker. May be mixed with other pesticides.

NAMES

***AVERMECTIN-B_1*, *ABAMECTIN*, AFFIRM, AGRIMEK, AVERT, AVID, DYNAMEC, MK-936, VERTIMEC, ZIMECTRIN**

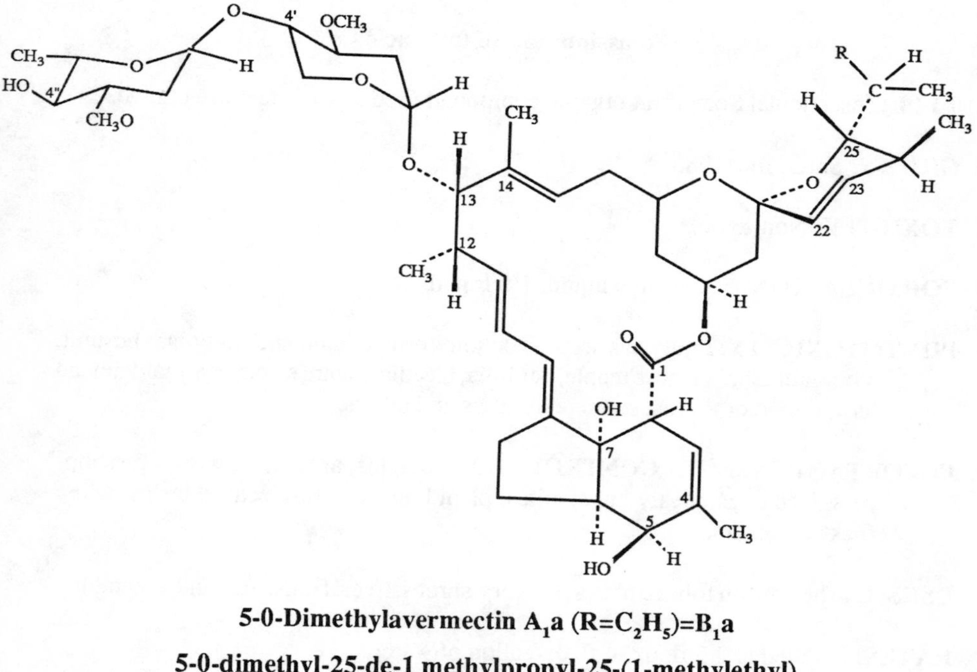

5-0-Dimethylavermectin A_1a ($R=C_2H_5$)=B_1a

5-0-dimethyl-25-de-1 methylpropyl-25-(1-methylethyl) avermectin A_1a ($R=CH_3$)=B_1b

(B_1a-80% B_1b-20%)

TYPE: A natural product containing macrocyclic lactone glycoside that is a product of fermentation used as an acaricide/insecticide.

ORIGIN: MSD Agvet, a Div. of Merck & Co., 1980.

TOXICITY: LD_{50}-10 mg/kg. May cause slight eye irritation.

FORMULATION: .011% bait, .15 lb ai EC.

PHYTOTOXICITY: Non-phytotoxic when used as directed. Do not use on ferns.

USES: Non-crop areas, ornamentals and turf. Experimentally being tested on ornamentals, citrus, cotton, vegetables, pears, potatoes, almonds, walnuts, apples, and other crops. Also being used to control imported fire ants. Used outside the U.S. on some of these crops.

IMPORTANT PESTS CONTROLLED: Mites, leafminers, pear psylla, fire ants, and others.

RATES: Applied at .005-.025 lb. a.i./A. (5-40 gm/ha)

APPLICATION: Apply when insects or mites appear and repeat as necessary, usually a 7-10 day schedule. To ants, apply when they are actively foraging.

PRECAUTIONS: Toxic to fish. Toxic to bees on contact. Do not apply if rainfall is expected.

ADDITIONAL INFORMATION: A stomach poison insecticide. This product is closely related to IVERMECTIN (MK-933) which is registered as a cattle parasiticide. A product of the fermentation by the soil microorganism *STREPTOMYCES AVERMITILIS*. This is a natural occurring product. Controls by both a contact and stomach poison activity. Slow-acting, although mites become immobilized after exposure. Good persistence and rainfastness on foliage. Active against mobil mites but no ovicidal activity. Minimum impact on beneficial insects. On fire ants it causes an immediate halt of egg production and the disappearance of the worker brood.

NAMES

THIOCYCLAM, EVISECT

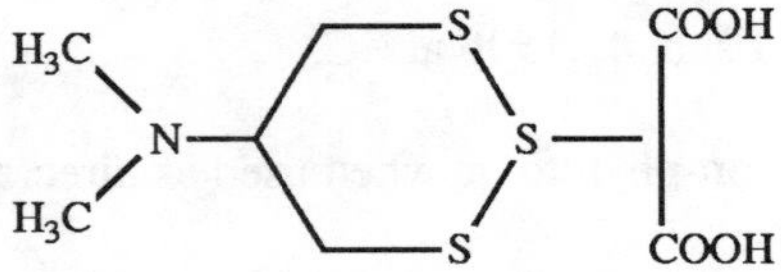

N,N-Dimethyl-1,2,3,-trithian-5-ylamine hydrogenoxalate

TYPE: Evisect is a nereistoxin-derivative insecticide with contact and stomach poison activity.

ORIGIN: Sandoz Ltd. (Basle, Switzerland), 1969.

TOXICITY: LD_{50}-310 mg/kg. May cause slight eye and skin irritation.

FORMULATIONS: 50% and 90% soluble powder, granules.

PHYTOTOXICITY: Safe for most crop plants when used at normal or double rates (exception Stark varieties of apples).

USES: Alfalfa, cabbage, corn, forestry, pears, potatoes, rape, sugarcane, tobacco, tomatoes, and others outside the U.S.

IMPORTANT PESTS CONTROLLED: Colorado potato beetle, leafminers, tuber worms, weevils, pear psylla, sawflies and others.

RATES: Applied at .300-1000 g/ha. kga.i./ha.

APPLICATION: Apply when insects appear, and repeat as necessary.

PRECAUTIONS: Not for sale or use in the U.S. Toxic to fish. Do not mix with copper compounds.

ADDITIONAL INFORMATION: Fungicidal activity has been observed against some rust fungi and stinking smut of wheat. Compatible with other pesticides.

NAMES

BUPROFEZIN, **APPLAUD, NNI-750**

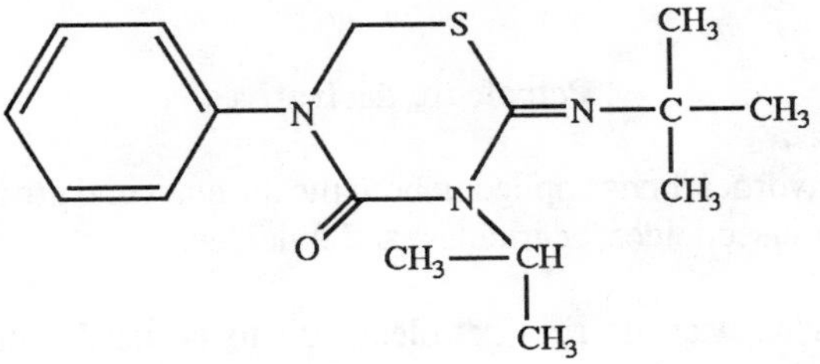

2-tert.-Butylimino-3-isopropyl-5-phenylperhydro-1,3,5-thiadiazin-4-one

TYPE: Applaud is a thiadiazine compound used as a contact and stomach-poison insecticide.

ORIGIN: Nihon Nohyaku of Japan, 1981.

TOXICITY: LD_{50}-2198 mg/kg.

FORMULATION: 10, 25, and 50% WP, 1.5% dust, 2% granule, 40% flowable.

PHYTOTOXICITY: Slightly phytotoxic to Chinese cabbage.

USES: Being used on rice, vegetable, fruit trees, and ornamental crops outside the U.S.

IMPORTANT INSECTS CONTROLLED: Planthoppers, mealybugs, leafhoppers, whiteflies, scales, Coleoptera, and others.

RATES: Applied at .125-1 kga.i./ha.

APPLICATION: Begin when insects appear and repeat as necessary.

PRECAUTIONS: Not for sale or use in the U.S. Adult insects are not controlled.

ADDITIONAL INFORMATION: Slow acting, but persists a long time. Low toxicity to mammals and fish, and no adverse effects on beneficial insects. Most effective against Hemiptera Species. Poor ovicidal activity. Control for up to 2 months has been observed. Temperatures have little effect on its activity. Good vapor action activity. Treated insects lay sterile eggs.

NAMES

DORMANT OILS, GB-1356, PETROLEUM OILS, SUMMER OILS, VOLCK OILS, WHITE OILS

Petroleum derivatives

TYPE: Oils are hydrocarbons applied in both the dormant and growing seasons, used as contact insecticides, acaricides, and ovicides.

ORIGIN: Kerosene was the first petroleum oil to be used around 1900. Higher distillations came into use around 1922. Volck oils by Chevron Chemical Company (Ortho Division of Standard Oil) first appeared in 1924.

TOXICITY: Relatively non-toxic.

FORMULATIONS:

1. Dormant oils-EC 85-90% oil with limited amounts of emulsifiers. Viscosity ranges from 90-150 seconds.
2. Summer oils-EC 60-97% actual oil, highly refined, with an unsulfonated residue to 90% or more. Viscosity ranges from 65-90 seconds.
3. Superior (Supreme) oils-New oils that increase plant safety, as well as insecticidal action.
4. Muscible oils (emulsifiable oils)-Water is added. The emulsifier is in the oil.
5. Stock emulsions-Contain up to 80% oil. The water is already added.

PHYTOTOXICITY: Dormant oils are extremely phytotoxic to all green plant parts. Summer oils are much safer. Various species and varieties vary in susceptibility, but tender, young foliage may be damaged in any case. Discoloration of blue spruce has resulted from its use. Weakened plants may show some damage.

USES:

1. Dormant oils-Used to control scales, aphid eggs, and mite eggs on almonds, apples, apricots, blueberries, caneberries, cashews, cherries, chestnuts, cranberries, currants, figs, filberts, gooseberries, grapes, huckleberries, nectarines, peaches, pears, pecans, plums, prunes, quinces, strawberries, walnuts, and ornamentals.
2. Petroleum solvents-Alfalfa, barley, buckwheat, clover, corn, cotton, flax, grasses, hops, lespedeza, millet, milo, oats, peanuts, popcorn, rice, rye, safflower, sorghum, soybeans, sugar beets, sugarcane, trefoil, vetch, wheat, cattle,

goats, poultry, sheep, swine, agricultural premises, and as a mosquito larvicide. Can be used as a space spray over stored grain.

3. Summer oils-Used to control aphids, mites, and scale crawlers on bush and vine fruits, citrus, pome fruits, stone fruits, grapes, olives, and vegetables.

4. Other uses-Herbicides on certain crops, parasiticides for application to livestock, and as carriers for other pesticides. Also used as a mosquito larvicide in aquatic areas.

IMPORTANT PESTS CONTROLLED:

1. Dormant oils-Scale, aphids, pear psylla, mites, leaf rollers, mealybugs, and others, as well as the eggs of many species.

2. Summer oils-Scale, mites, mealybugs, whitefly, aphids, pear psylla, and others, as well as the eggs of many species.

RATES:

1. Dormant oils-Applied at 30-250 gal actual/A.
2. Summer oils-Applied at 15-400 gal actual/A.
3. Mosquito Larvicide-Use 1-3 gal/A.

APPLICATION:

1. Dormant oils-Diluted sprays containing 2-7% oil are applied to fruit trees, bush fruits, and ornamentals when they are in the dormant period. Apply when the temperature is above 40°F. They are phytotoxic to foliage.

2. Summer oils-Diluted sprays containing .24-2% oils are used. Often mixed with other pesticides. There are no limitations on the application time during the growing season. They are much less phytotoxic than the dormant oils.

3. Mosquito Larvicide-Apply by ground or air to aquatic areas to control larvae and pupae.

PRECAUTIONS: Sulfur sprays shouldn't follow an oil spray for 2-3 weeks. Injury to rubber hose and rubber parts of spray equipment may result. Do not apply within 60 days after or 90 days before applying Captan or Phaltan. Do not combine with Sevin. Do not apply when the temperature is above 90°F or below 35°F.

ADDITIONAL INFORMATION: On foliage, a refined oil of narrow viscosity range and a high content of unsulfonated residue is advisable (sulfur residue on trees will make such oils phytotoxic.) The lower the viscosity of the oils, the safer they are to use on plants. The smaller the amount of unsaturated hydrocarbons present in an oil, the safer it is to use on plants. Considered to be of rather low toxicity to insects. However, no insect resistance to oils has developed.

RELATED COMPOUNDS:

1. Elgetol (DNOC) — An older insecticide produced by FMC to use on apples to control scale, aphids, and apple scab. Also used to control diseases in mushroom houses and for the blossom thinning of apples.

NAME

AROSURF-MSF

Poly (oxy-1,2-ethanediyl), alpha-isooctalecyl-omega-hydroxy

TYPE: Asosurf-MSF is a non-petroleum surface active oil used as a mosquito larvicide and pupicide.

ORIGIN: Sherex Chemical Co., 1982.

TOXICITY: LD_{50}-20,000 mg/kg. May cause skin irritation.

FORMULATION: 100% active liquid.

PHYTOTOXICITY: Not to be sprayed directly on plant foliage.

IMPORTANT PESTS CONTROLLED: Mosquitoes.

USES: Used on water surfaces.

RATES: Applied at .2-.5 gal/surface A.

APPLICATION: Applied by ground or air over the water surface. Spray systems should be equipped with high sheer agitation. Repeat as necessary.

PRECAUTIONS: Wind, run-off, etc., may result in poor control due to displacement of the water.

ADDITIONAL INFORMATION: Not visible on the water surface. Forms a film over the water surface that prevents the mosquito larvae from breathing. Larvae kill will usually occur in 24-72 hours and pupicidal activity in 24 hours.

NAMES

***FENVALERATE*, AQMATRINE, BELMARK, ECTRIN, EXTRIN, FENVALETHRIN, MOSCADE, PYDRIN, PYRID, SAN MARTON, SUMIBAC, SUMICIDIN, SUMIFLEECE, SUMIFLY, SUMITICK, TIRADE, TRIBUTE**

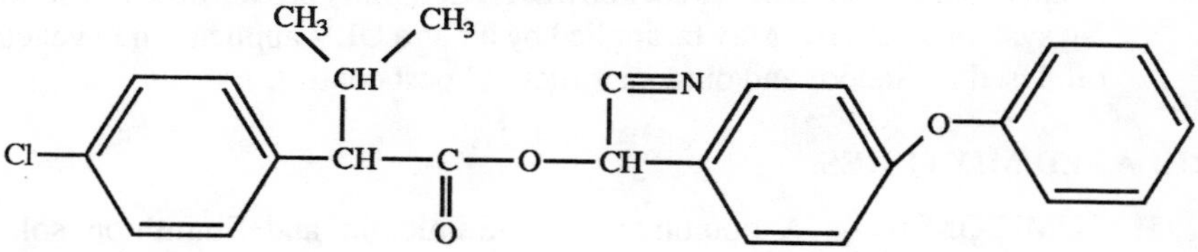

(s)-Cyano(3-phenoxyphenyl)methyl 4-chloro alpha (1-methylethyl)benzeneacetate

TYPE: Pydrin is a synthetic-pyrethroid compound used as selective contact and stomach-poison insecticide.

ORIGIN: Sumitomo Chemical Co. of Japan, 1974. Being developed and sold in the U.S. by DuPont and in other areas by Shell Chemical Co.

FORMULATIONS: 2.4 EC, 10% WDL, 10, 20, and 30% EC.

TOXICITY: LD_{50}-451 mg/kg. May be irritating to the eyes and skin.

PHYTOTOXICITY: Non-phytotoxic when used as recommended. Spotting has occurred on tomatoes.

USES: Agricultural premises, almonds, apples, apricots, artichokes, beans, blueberries, broccoli, cabbage, caneberries, carrots, cauliflower, cherries, collards, corn, cotton, eggplant, gooseberries, filberts, grapes, huckleberries, lettuce, livestock and livestock premises, melons, nectarines, okra, ornamentals, peaches, peanuts, pears, peas, pecans, peppers, plums, potatoes, prunes, pumpkins, radishes, soybeans, squash, sugarcane, sunflowers, and tomatoes. Also used on noncropland to control grasshoppers and other insects and as a crack and crevice cockroach spray. Can be used in food handling facilities.

IMPORTANT PESTS CONTROLLED: Cotton armyworms, cutworms, corn borer, flies, cockroaches, crickets, fleas, earwigs, spiders, pear psylla, bollworm, cotton leaf perforator, leafhoppers, thrips, tobacco budworm, lygus, and many other lepidoptera species. Aphids are controlled by direct spray.

RATES: Applied at .05-.2 lba.i./A.

APPLICATION: Apply when insects appear, and repeat as necessary.

PRECAUTIONS: Toxic to fish and bees. Do not graze the treated areas. Do not mix with alkaline materials.

ADDITIONAL INFORMATION: May be applied by ground or air. Gives good residual control with a fast knockdown. Is not rapidly broken down by sunlight. No systemic activity. May be applied by air as a ULV application in vegetable oil. Used for indoor and outdoor structural pest control.

RELATED MIXTURES:

1. SUMICOMBI — A combination of Sumicidin and Sumithion sold by Sumitomo Chem. Co. in Japan.

NAMES

ESFENVALERATE, SUMI-alpha, ASANA, HALMARK

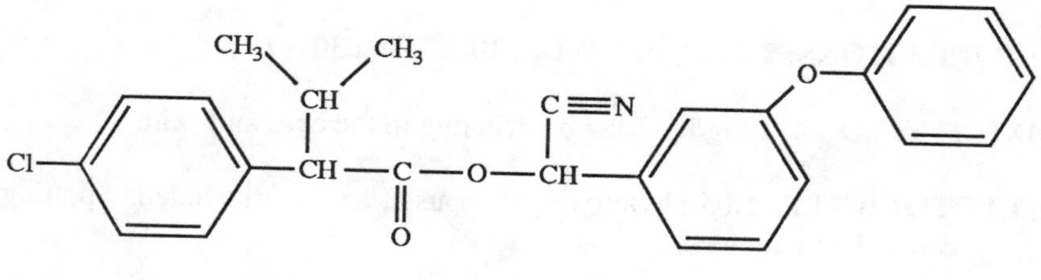

CH_3 CH_3 CH C≡N Cl CH C O CH O O

(S)-alpha-Cyano-3-phenoxybenzyl (S)-2-(4-chlorophenyl)-s-methylbutyrate

TYPE: Esfenvalerate is a synthetic pyrethroid compound which is the alpha isomer of fenvalerate.

ORGIN: 1982 Sumitomo Chemical Co. Also being developed in certain countries by Shell Chemical Co. and DuPont Co.

TOXICITY: LD_{50} 325 mg./kg.

FORMULATIONS: 25 EC, 50 EC.

PHYTOTOXICITY: Some injury has been noted on crucifers, cucumbers, eggplant, tomatoes, pears and mandrin oranges.

USES: Corn, potatoes, cotton, wheat, rape, hops, apples, pears, grapes, peaches, cucurbits, peppers, tomatoes, crucifers and other crops as Pydrin.

IMPORTANT INSECTS CONTROLLED: European corn borer, Colorado potato beetle, bollworms, aphids, white fly, loopers, codling moth, pear psylla, leafminer and others.

RATES: Applied at 7.5-50 g. a.i./ha.

APPLICATION: Apply when insects appear and repeat as necessary.

PRECAUTIONS: Toxic to fish. Do not mix with alkaline materials.

ADDITIONAL INFORMATION: Up to 4 times more active than fenvalerate. Controls as a contact or stomach poison insecticide. Also gives aphid control.

NAMES

***PERMETHRIN*, AMBUSH, ATROBAN, BIOTHRIN, BRUNOL, COOPEX, CORSAIR, DETMOL, DRAGNET, ECTIBAN, EKSMIN, EVERCIDE, GARD-STAR, HARD-HITTER, IMPERATOR, INSECTABAN, KAFIL, OUTFLANK, OVER-TIME, PERIGEN, PERMANDINE, PERMECTRIN, PERMIT, PERTHRINE, POUNCE, PRAMEX, QAMLIN, RESIDROID, RONDO, STOCKADE, STOMOXIN, TALCORD, TORNADE, TORPEDO**

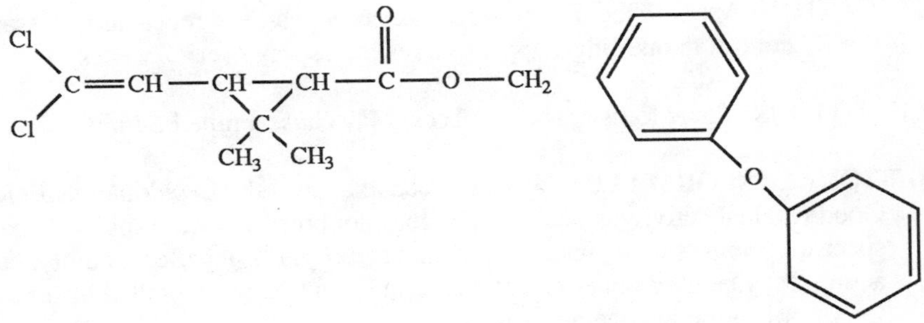

(3-Phenoxyphenyl)-methyl(±)cis-trans-3-(2,2-dichloroethenyl)-2,2) dimethylcyclopropane-carboxylate

TYPE: Permethrin is a synthetic-pyrethroid compound used as a contact and stomach-poison pesticide.

ORIGIN: National Research Development Corp., England, 1974. Licensed to be developed and manufactured in the U.S. by FMC Corp. and ICI Americas. Sold outside the U.S. by Shell Chemical Co. also.

TOXICITY: LD_{50}-450 mg/kg. Causes some eye and skin irritation.

FORMULATIONS: 3.2 EC, 25% WP, 1.5 Granules 2 EC, 1 EC, 4 EC. Impregnated in ear tags for use on livestock.

PHYTOTOXICITY: Non-phytotoxic when used as directed. Injury has occurred on certain ornamentals.

USES: Agricultural premises, alfalfa, almond, apples, artichokes, asparagus, broccoli, Brussels sprouts, cabbage, cauliflower, celery, cherries, chrysanthemums, collards, corn, cotton, filberts, horseradish, lettuce, livestock, melons, mushroom houses, peaches, pears, pistachios, potatoes, pumpkins, rangeland, soybeans, spinach, turnips and walnuts. Use on these crops outside of the U.S., as well as a livestock and public health insecticide.

RATES: Applied at .05-.2 lba.i./A.

IMPORTANT PESTS CONTROLLED: Alfalfa weevil, armyworms, boll weevils, bollworms, budworms, cabbageworms (imported), codling moth, Colorado potato beetle, cotton leaf perforator, diamondback moth, European corn borer, flea beetles, flies, houseflies, leaf miners, lice, loopers, lygus, Mexican bean beetle, mosquitoes, naval orangeworm, peach twig borer, pear psylla, termites, ticks, whiteflies, and many others.

APPLICATION: Apply to the foliage when insects appear, and repeat as necessary. For fly control in buildings, apply as a residual spray every 2 weeks.

PRECAUTIONS: Toxic to fish. Toxic to bees. May cause a mite buildup.

ADDITIONAL INFORMATION: Most effective against the Lepidoptera species. Good residual activity, since sunlight does not break it down rapidly. Broad-spectrum. Nonsystemic, with no fumigant activity. Has repellent activity. Fast acting. May be used under greenhouse conditions. May be applied by air as a ULV application in vegetable oil.

NAMES

CYPERMETHRIN, AGROTHRIN, AMMO, ARRIVO, AVICADE, BARRICIDE, CYMBUSH, CYMPERATOR, CYNOFF, CYPERKILL, DEMON, ECTOROR, EKTOMIN, EQUIBAND, FENOM, FLECTRON, FOLCARD, KAFIL SUPER, KRUEL, NURELLE, PARZON, POLYTRIN, RIPCORD, SHERPA, STOCKADE, TOPPEL

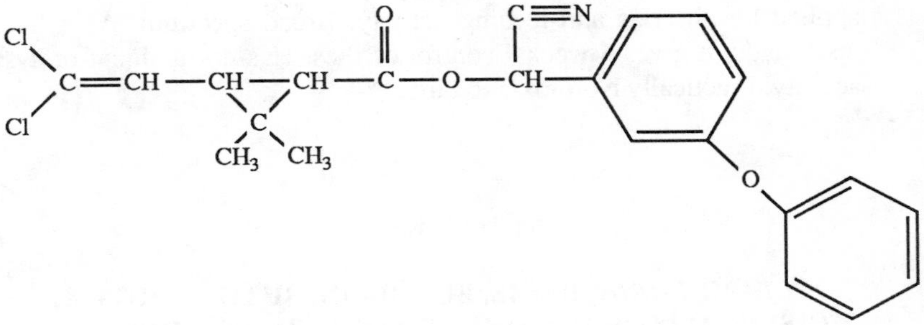

(±) alpha-Cyano-3-phenoxybenzyl (±) cis. trans 3-(2,2-dichlorovinyl)-2,2-dimethyl cyclopropane-carboxylate

TYPE: Cypermethrin is a synthetic-pyrethroid compound used as a contact insecticide that also has stomach-poison activity.

ORIGIN: ICI, FMC, CIBA-Geigy, Sumitomo, and Shell Chemical Co., 1975.

TOXICITY: LD_{50}-200 mg/kg. May cause skin and eye irritation.

FORMULATIONS: 5%, 2.5 EC, 3 EC, 1.5% ULV, and others.

PHYTOTOXICITY: Non-phytotoxic when used as directed.

IMPORTANT INSECTS CONTROLLED: Lepidoptera spp., Coleptera spp., Diptera, Hemiptera, Homoptera spp., and many others.

USES: Cotton, pecans, and lettuce. Sold in many countries for use on cotton, and as a livestock insecticide. Either being sold or developed in certain countries for use on cotton, tobacco, citrus, fruit tress, vines, corn, cocoa, coffee, soybeans, rice, vegetables, sugar beets, ornamentals, and other crops. Sold impregnated in ear tags for use on livestock in some countries. Used as a crack and crevise treatment by pest control operators in non-food areas.

RATES: Applied at .02-.075 lba.i./A.

APPLICATION: Apply when insects appear, and repeat as necessary. Good coverage is necessary, since it has no systemic activity. Use on livestock as a spray. Remains biologically active on the plant longer than conventional insecticides.

PRECAUTIONS: Toxic to fish and other aquatic life. Toxic to bees. Not effective as a soil insecticide.

ADDITIONAL INFORMATION: Slightly more active than permethrin. Aphids will be controlled when directly sprayed. Fast knockdown activity. May be applied by air. Has anti-feeding activity. Broad spectrum. As a livestock insecticide, it gives 4 weeks' control of flies. Has no fumigant or systemic activity. Practically Non-toxic to birds.

NAMES

***DELTAMETHRIN,* BECIS, BUTOFLIN, BUTOSS, BUTOX, CISLIN, CRACK DOWN, DECIS, K-OTHRIN, NRDC-161**

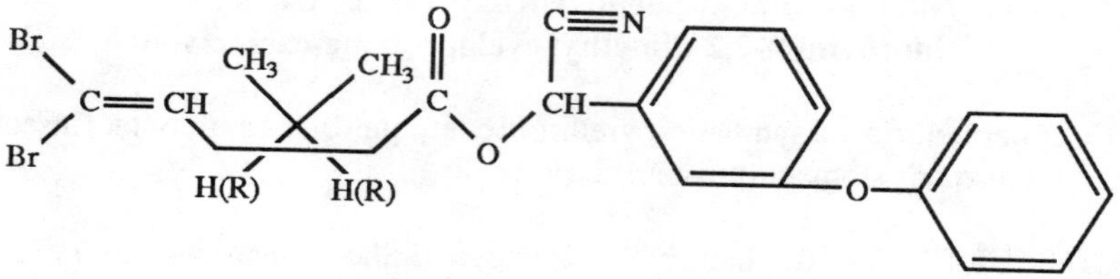

(s)-alpha-Cyano-M-phenoxybenzyl (1*R*, 3*R*)-3-(2,2-dibromovinyl)-2,2-dimethyl cyclopropane-carboxylate

TYPE: Decamethrin is a synthetic-pyrethroid compound used as a contact and stomach-poison insecticide.

ORIGIN: Roussel Uclof of France, 1975. Being developed by Procida of France.

TOXICITY: LD_{50}-135 mg/kg.

FORMULATIONS: ULV, EC, WP, dust, or granules.

PHYTOTOXICITY: Non-phytotoxic when used as directed.

USES: Used to protect cereals and legumes against stored grain insects. Widely used outside the U.S. on numerous crops. Being sold in some countries as a public health insecticide.

RATES: Applied at 7.5-25 gramsa.i./ha.

IMPORTANT PESTS CONTROLLED: Coleoptera, houseflies, cockroaches, flies, mosquitos, stroed grain insects, lepidoptera, thrips, and many other insects.

APPLICATION: Apply when insects appear, and repeat as necessary. Will give 7-21 days' control. Used to treat grain in storage.

PRECAUTIONS: Not for sale or use in the U.S. Toxic to fish and other aquatic life. Toxic to bees.

ADDITIONAL INFORMATION: A stable compound when exposed to air or light. Very broad spectrum.1000 times as toxic to houseflies as pyrethrins. Considered the most powerful of the synthetic pyrethroids. Fast acting. Some repellent activity. No systemic activity. Used as a residual treatment for cockroaches, etc.

NAMES

***FENPROPATHRIN,* DANITOL HERALD, KILUMAL,MEOTHRIN, RANDAL, RODY, S-3206, SMASH, SUMIRODY**

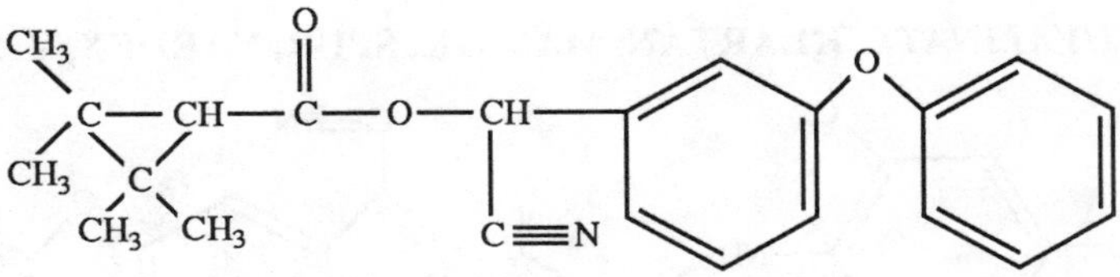

alpha-Cyano-3-phenoxybenzyl 2,2,3,3,-tetramethyl cyclopropanecarboxylate

TYPE: Danitol is a synthetic-pyrethroid compound used as a selective insecticide-acaricide with repellancy and contact activity.

ORIGIN: Sumitomo Chemical Co., Ltd., 1973. Being developed by Shell Int'l. in some countries and by Valent Chemical Co. in the U.S.

TOXICITY: LD_{50} -71.8 mg/kg. May cause eye and skin irritation.

FORMULATIONS: 2.4 lb/gal. EC, and 5% WP.

PHYTOTOXICITY: Non-phytotoxic when used as directed.

IMPORTANT PESTS CONTROLLED: Spider mites, thrips, codling moth,whiteflies, leaf miners, bollworms, leafworms, leaf rollers, armyworms, loopers, aphids, tortrixes, psylla, tuberworms, cutworms, mosquitoes, and many others.

USES: Experimentally used on fruit trees, turf, cotton, grapes, citrus, ornamentals, vegetables, flowers, and field crops. Used on these outside the U.S.

RATES: Applied at .1-.4 lb. ai in 30-400 gallons of water.

APPLICATION: Apply when insects appear, and repeat as necessary.

PRECAUTIONS: Toxic to fish. Used on an experimental basis only.

ADDITIONAL INFORMATION: Highly effective on various species of mites, but not on rust mites. More effective at lower temperatures. Primary action is its strong repellancy. Contact activity against nymphs and adults, with ovicidal activity. Residual effect is mainly due to its strong repellancy. No vapor activity.

NAMES

***FLUVALINATE,* KLARTAN, MAVRIK, SPUR, YARDEX, ZR-3210**

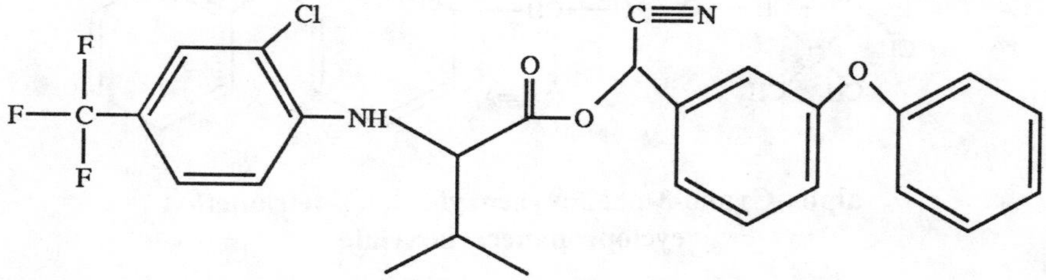

(RS)-alpha-Cyano-3-phenoxybenzyl(R)-2-[2-chloro-4-(trifluoromethyl)anilino]-3-methylbutanoate

TYPE: Mavrik is a synthetic-pyrethroid compound used as a selective contact and stomach-poison insecticide.

ORIGIN: Sandoz Crop Protection, 1977.

TOXICITY: LD_{50}-261 mg/hg. May cause eye and skin irritation.

FORMULATION: 2 EC, 2 lb/gal flowable. 25% WP.

PHYTOTOXICITY: Non-phytotoxic when used as directed.

USES: Ornamentals, cotton, and nonbearing tree and vine crops, and non-crop areas. Experimentally being tested on cole crops, lettuce, alfalfa, tobacco, tomatoes, corn, and other crops. Used on a number of crops outside the U.S. and on crops grown for seed.

IMPORTANT PESTS CONTROLLED: Budworms, bollworms, boll weevils, thrips, mites, whiteflies, cotton leaf perforator, lygus, loopers, earworms, armyworms, aphids, and others.

RATES: Applied at .025-.1 lb actual/A.

APPLICATION: Apply when insects appear, and repeat as necessary, usually on a 5-10 day basis. May be used inside greenhouses.

PRECAUTIONS: Do not use in fogging type applicators. Toxic to fish.

ADDITIONAL INFORMATION: Suppresses spider mite populations. Maintains its activity under high-temperature conditions. May be tank mixed with other products.

NAMES

FLUFENOXURON, WL 115110, CASCADE

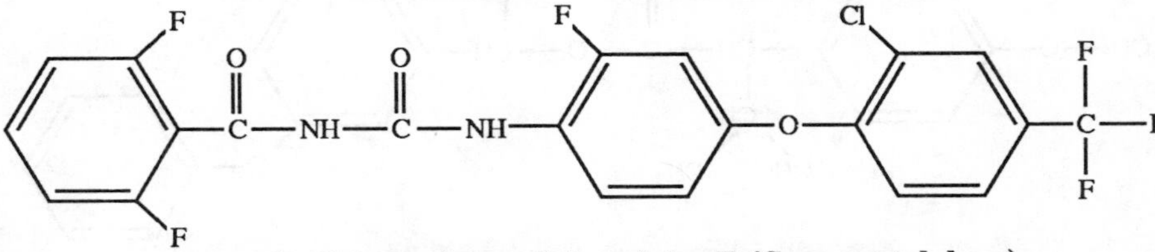

1-[4-(2-Chloro-alpha, alpha, alpha-Trifluoro-*p*-tolyloxy)-2-fluorophenyl]-3-(2,6-difluorobenzoyl)urea

TYPE: Flufenoxuron is an acylarea compound used as a broad spectrum insecticide and miticide.

ORIGIN: Shell Chemical Co, in England-1982.

TOXICITY: LD_{50} 3000 mg/kg.

FORMULATION: 5% WDC.

PHYTOTOXICITY: Non-phytotoxic when used as directed.

USES: Experimentally being tested on cotton, corn, coffee, grapes, orchard crops, citrus, vegetable and others.

IMPORTANT INSECTS CONTROLLED: Mites, psyllids, fruit tortrix, leafminers, grape berry moth, scales, diamondback moth and many others.

RATES: Applied at 5-40 grams a.i./ha.

APPLICATION: Apply when insects appear and repeat as necessary.

PRECAUTIONS: Used on an experimental basis only.

ADDITIONAL INFORMATION: A wide spectrum product. Early season application gives the best control. Does not control adult insects, only the larvae. Adult insects that are treated do not deposit viable eggs. Gives long lasting control. Fast acting.

NAMES

***FLUCYTHRINATE,* AC 22705, CYBOLT, GUARDIAN, PAY-OFF**

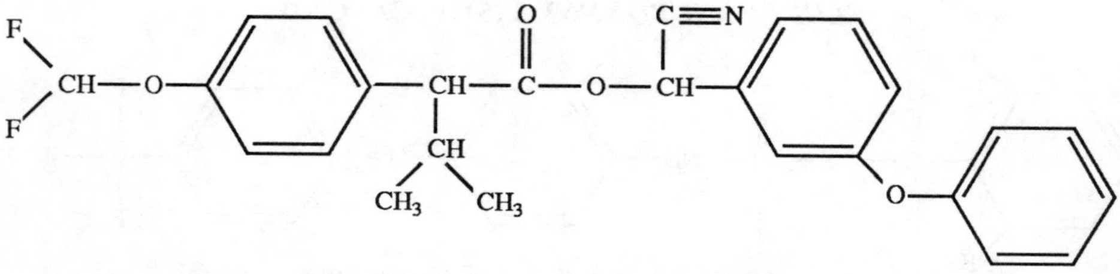

(±)-Cyano (3-phenoxyphenyl) methyl(±)-4-(difluromethoxy) alpha-(1-methylethyl) benzene aceate

TYPE: Pay-Off is a synthetic-pyrethroid compound with a broad spectrum of activity, used as a contact and stomach-poison insecticide.

ORIGIN: DuPont Co., 1979.

TOXICITY: LD_{50}-67 mg/kg. Irritating to the eyes.

FORMULATIONS: 2.5 EC.

PHYTOTOXICITY: Non-phytotoxic when used at the suggested rates.

USES: Cotton, corn, cabbage, pears, cattle, and lettuce, Experimentally being tested on corn, vegetable, and fruit crops.

IMPORTANT PESTS CONTROLLED: Bollworms, budworms, loopers, saltmarsh caterpillar, aphids, plant bugs, whiteflies, and many others.

RATES: Applied at .025-.08 lb a.i./A.

PRECAUTIONS: Toxic to fish and other aquatic organisms. Toxic to bees.

ADDITIONAL INFORMATION: Very effective in controlling the Heliothis complex of insects. No fumigant or systemic activity. May be applied by air. Also used as a ULV spray in vegetable oil. Sold impregnated in ear tags for use on cattle for fly control.

RELATED MIXTURES:

1. AASTAR — A combination of phorate and flucythrinate used on field corn as a soil systemic insecticide. Developed by American Cyanamid.

NAMES

CYFLUTHRIN, BAY-FCR-1272, BAYTHROID, LASER, SOLFAC, TEMPO

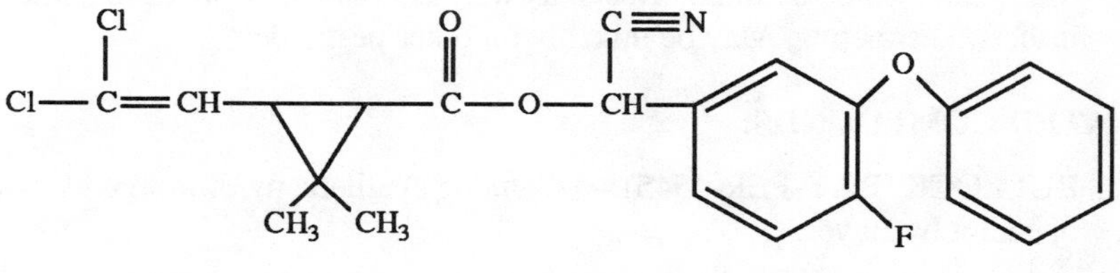

Cyano(4-fluoro-3-phenoxyphenyl) methyl 3-(2,2-dichloroethenyl)-2,2-dimethyl-cyclopropanecarboxylate

TYPE: BAY-FCR 1272 is a synthetic-pyrethroid compound being used as a contact and stomach-poison insecticide.

ORIGIN: Bayer AG of West Germany, 1980. Being developed in the U.S. by Mobay Chemical Corp.

TOXICITY: LD_{50}-291 mg/kg. May cause eye irritation.

FORMULATIONS: 2 EC, 20% WP. ULV formulations. 10% WP.

PHYTOTOXICITY: Non-phytotoxic when used as recommended.

USES: Cotton and ornamentals. Also used as a public health insecticide. Experimentally in the U.S. being tested on alfalfa, rice, hops, sugar beets, corn, peanuts, potatoes, soybeans, tobacco, apples, grapes, pears, cole crops, lettuce, tomatoes, and other crops. Also as a household, stored products, and public health insecticide. Being sold for many of these purposes outside the U.S.

IMPORTANT PESTS CONTROLLED: Cutworms, silverfish, cockroaches, termites, grain beetles, weevils, mosquitoes, fleas, flies, ants, corn earworms, tobacco budworm, codling moth, European corn borer, cabbageworm, loopers, armyworms, boll weevil, alfalfa weevil, Colorado potato beetle, and many others.

RATES: Applied at .0125-.1 lb kg a.i./acre.

APPLICATION: Apply when insects appear and repeat as necessary. Applied as a crack and crevise treatment on household pests.

PRECAUTIONS: Not effective against sub-surface soil insects. Toxic to fish. Toxic to bees. Do not tank mix with PEROPAL.

ADDITIONAL INFORMATION: Nonsystemic. Very effective against chewing insects at lower rates than the presently used synthetic pyrethoids. Shows good side effects against sucking insects, provided they are contacted by the spray. Many have some ovicidal effects, as well as being a repellent to some adult insects. Fast-acting. May be mixed with other pesticides.

RELATED COMPOUNDS:

1. BULLOCK (BAY-FER-4545) — A analog synthetic pyrethroid under-developement by Bayer.

NAMES

***ALPHACYPERMETHRIN*, BESTOX, DOMINEX, FASTAC, FENDONA, WL-85871**

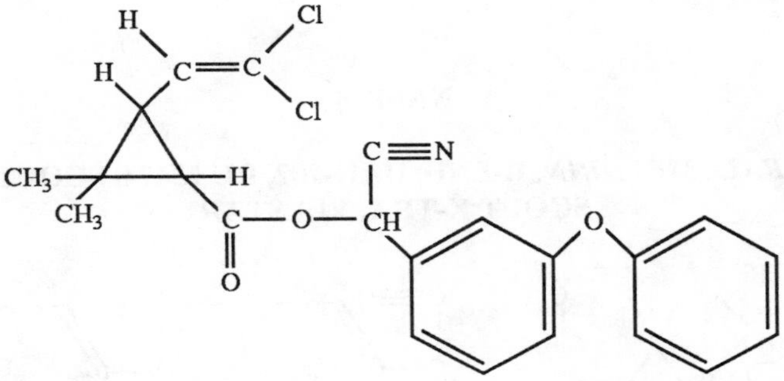

(1R cis) S and (1S cis) R enantiomer isomer pair of alpha-Cyano-3-phenoxybenzyl-3-(2,2-dichlorovinyl)-2, 2-dimethylcyclo-propane carboxylate

TYPE: FASTAC is a synthetic pyrethroid compound used as a contact and stomach-poison insecticide.

ORIGIN: Shell Research Ltd. of England, 1982.

TOXICITY: LD_{50}-79 mg/kg. May cause eye and skin irritation.

FORMULATIONS: 50 g. 100 g and 200 g/1 EC, 50 g/kg WP.

PHYTOTOXICITY: Non-phytotoxic when used as directed.

USES: Used outside the U.S. on cereals, citrus, coffee, cotton, forestry crops, flower crops, fruit crops, potatoes, rape, rice, soybeans, sugar beets, tea, tobacco, vegetables, and other crops. Also used as a public health insecticide.

IMPORTANT PESTS CONTROLLED: Weevils, beetles, cutworms, flies, mosquitos, cockroaches, leaf miners, fruit flies, maggots, aphids, scales, whiteflies, leafhoppers, mealybugs, leaf rollers, bollworms, cutworms, armyworms, loopers, grasshoppers, thrips, and others.

RATES: Applied at 5-20 g a.i./ha.

USES: Apply when insects appear and repeat as necessary.

PRECAUTIONS: Not for sale or use in the U.S. Mites are not controlled. Toxic to bees and fish.

ADDITIONAL INFORMATION: Fast acting. Has ovicidal activity. Excellent stability on the plant. Resistant to rain wash off since it is virtually insoluable in water.

NAMES

TRALOMETHRIN, D-END, HAG-107, RU 25474, SCOUT, SCOUT X-TRA, VELSTAR

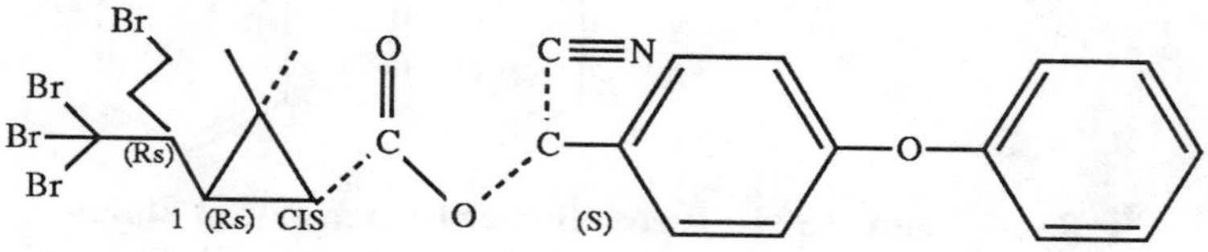

Br Br Br Br (Rs) 1 (Rs) CIS O C O C≡N C (S) O

[1R,3S)3[(1'RS)(1',2',2',2'-Tetrabromoethyl)]-2,2-dimethyl cyclopropane carboxylic acid (s)-alpha-cyano-3-phenoxybenzyl ester

TYPE: Scout is a synthetic pyrethroid compound used as a contact and stomach-poison insecticide.

ORIGIN: Roussel-Ucalf of France, 1979. Being developed in the U.S. by Hoechst.

TOXICITY: LD_{50}-99.2 mg/kg. May cause eye and skin irritation.

FORMULATIONS: .3 lb/gal EC.

PHYTOTOXICITY: Non-phytotoxic when used as directed.

USES: Cotton and soybeans. Experimentally being tested on corn, alfalfa, vegetable crops, tomatoes, potatoes, pears, and many other crops.

IMPORTANT PESTS CONTROLLED: Cotton leaf perforator, Mexican bean beetle, green clover worms, armyworms, tobacco budworm, bollworm, loopers, armyworms, cutworms, thrips, European corn borer, leafhoppers, grasshoppers, cabbageworm, potato tuberworm, corn earworm, pear psylla, and many others.

RATES: Applied at .013-.024 lb a.i./A.

APPLICATION: Apply when insects appear and repeat as necessary. Apply by ground or air.

PRECAUTIONS: Corrosive to some metals. Toxic to fish.

ADDITIONAL INFORMATION: May be used with vegetable oil when applied in an ULV program. Will suppress the population of bollweevils, but not control them. No toxicity to bees was noted when they contacted treated foliage an hour after application. Applied as a crack and crevice treatment for household peasts.

NAMES

***CYCLOPROTHRIN*, BACLASH, CYCLOSAL, NKI-8116**

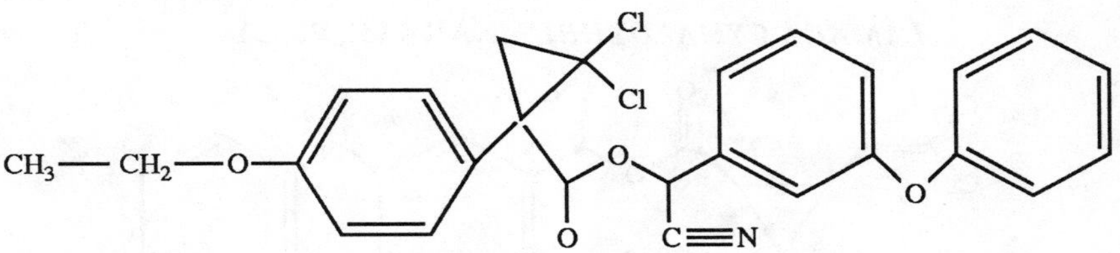

alpha-Cyano-3-phenoxy benzyl 1-(p-ethoxyphenyl)-2, 2-dichloro-cyclopropanecarboxylate

TYPE: Cycloprothrin is a synthetic-pyrethroid compound used as a contact and stomach-poison insecticide with anti-feeding and repellent activity.

ORIGIN: CSIRO of Australia, 1981. Being developed by R. Maag of Australia and Nippon Kayaku of Japan.

TOXICITY: LD_{50}-5000 mg/kg.

FORMULATIONS: 15 and 25% WP, 10, 15 and 20% EC, .5-1% dust, 2% granules. 65% premix.

PHYTOTOXICITY: Non-phytotoxic when used as directed.

USES: Outside the U.S. on cotton, potatoes, fruit crops, soybeans, vegetables, citrus, tea, rice, and others. Also used on sheep to control sheep blowfly.

IMPORTANT INSECTS CONTROLLED: Leaf rollers, leaf miners, oriental fruit moth, cabbageworms, armyworms, cutworms, loopers, rice water weevil, thrips, stinkbugs, aphids, and others.

RATES: Applied at .01-.2 kg a.i./ha.

APPLICATION: Apply when insects appear and repeat as necessary. The dust and granule formulations are to be used on rice.

PRECAUTIONS: Not for sale or use in the U.S.

ADDITIONAL INFORMATION: Low fish toxicity. Killing efficiency is no better than organophosphates, but it shows superior control in the field due to flushing out, repellent activity, and antifeeding activity. Compatable with other insecticides. Used as a livestock insecticide in Australia.

NAMES

***LAMBDA CYHALOTHRIN*, KARATE, PP321**

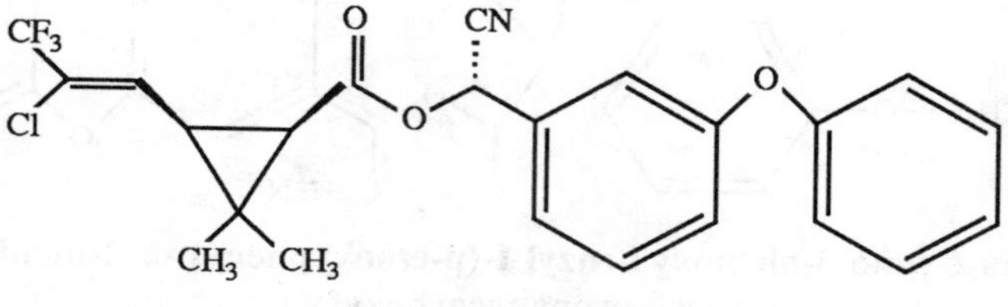

CF_3 O CN Cl O O CH_3 CH_3

and enantiomer (1: 1 mixture)

[1 alpha (s), 3 alpha(z)]-(±)-Cyano(3-phenoxybenzyl) methyl-3- (2-chloro-3,3,3-trifluoro-1-propenyl)-2,2-dimethylcyclopropane carboxylate

TYPE: Karate is a synthetic-pyrethroid contact and stomach-poison insecticide.

ORIGIN: ICI of England, 1982.

TOXICITY: LD_{50}-56 mg/kg. May cause eye irritation.

FORMULATIONS: 25 g/1 EC, 50 g/l EC, 8 g/1 UL.1

PHYTOTOXICITY: Non-phytotoxic when used as directed.

USES: Cotton, Experimentally being tested on fruit, vegetables, and field crops and for public health pest control.

IMPORTANT PESTS CONTROLLED: Aphids, lygus, Colorado potato beetle, boll weevil, bollworms, budworms, armyworms, leaf miners, codling moth, mites, cockroaches, houseflies, mosquitoes, whitefly, leaf hooper, cutworms, cabbage looper, European corn borer, thrips, and many others.

RATES: Applied at 5-30 g a.i./ha.

APPLICATION: Apply when insects appear and repeat as necessary.

PRECAUTIONS: It can not be soil incorporated. Toxic to fish.

ADDITIONAL INFORMATION: Compatible with most insecticides and fungicides. Rain fast after the spray deposit has dried. Gives both rapid knockdown and persistent control. Suppresses white flies and mites.

NAMES

***TEFLUTHRIN*, FORCE, PP-993**

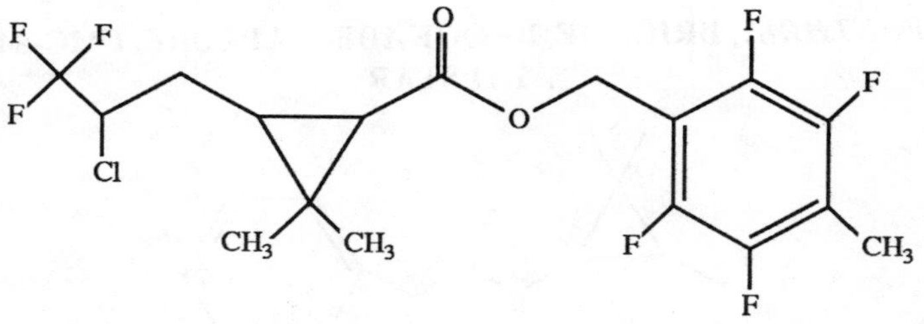

2,3,5,6-Tetrafluoro-4-methylbenzyl (Z)-(1*RS*,3*RS*)-3-(2-chloro-3,3,3-trifluoroprop-1-ethyl)-2,2-dimethylcyclopropane carboxylate

TYPE: Tefluthrin is a synthetic pyrethroid compound being used as a soil insecticide.

ORIGIN: ICI of England 1983.

TOXICITY: LD_{50} 22 mg/kg. May cause slight eye and skin irritation.

FORMULATION: 1.5% granules.

PHYTOTOXICITY: Non-phytotoxic when used as directed.

USES: Corn. Experimentally being tested on sorghum, small grains, potatoes, vegetables and other crops.

IMPORTANT INSECTS CONTROLLED: Corn rootworms, wire worms, flea beetle, chafer, cutworms, corn borer, fruit fly, wheat bulb fly, symphylids and others.

RATES: Applied at 12-150 g. a.i./ha.

APPLICATION: Applied as a soil granule, a liquid soil or foliar spray or as a seed treatment.

PRECAUTIONS: High fish toxicity.

ADDITIONAL INFORMATION: Broad spectrum. Earthworms are not harmed. The half life in the soil is 1-3 months, giving long lasting control. The first systemic pyrethroid designed as a soil insecticide.

NAMES

***BIFENTHRIN,* BRIGADE, BROOKADE, CAPTURE, FMC-54800, TALSTAR**

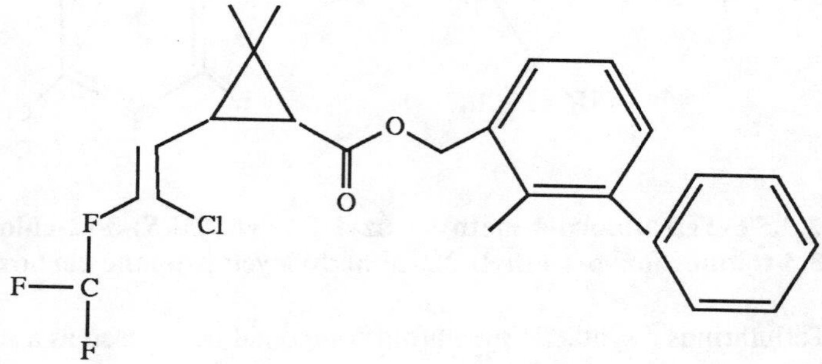

[1 alpha, 3 alpha(z)]-(+ or -)-(2-methyl[1,1'-biphenyl]-3-yl) methyl 3-(2-chloro-3,3,3-trifluoro-1-propenyl)-2,2-dimethylcyclopropane-carboxylate

TYPE: Biphenate is a synthetic-pyrethroid compound used as a contact and stomach-poison insecticide/acaricide.

ORIGIN: FMC Corp., 1983.

TOXICITY: LD_{50}-54.5 mg/kg. Causes eye and skin irritation.

FORMULATIONS: 2 EC, 10% WP, 100 g/L EC, 80 g/L flowable.

PHYTOTOXICITY: Non-phytotoxic when used as directed. Do not apply to poinsettias, primrose, marigold, crape myrtle, gerber daisy and ferns.

USES: Ornamentals and cotton. Experimentally being tested on alfalfa, almonds, apples, corn, cotton, citrus, grapes, pears, peaches, potatoes, soybeans, strawberries, and other crops.

IMPORTANT PESTS CONTROLLED: Mites, alfalfa weevil, aphids, leafhopper, peach twig borer, codling moth, plum curculio, leaf rollers, scales, corn earworm, armyworm, bollworm, budworm, boll weevil, loopers, lygus, whiteflies, thrips, pear psylla, Mexican bean beetle, Colorado potato beetle, and others.

RATES: Applied at .04-.2 lb a.i./A.

APPLICATION: Apply when insects and mites appear and repeat as necessary.

PRECAUTIONS: To be used on an experimental basis only. Toxic to fish. Toxic to bees.

ADDITIONAL INFORMATION: Effective for the control of mites. No systemic activity.

NAMES

***KADETHRIN*, SPRAY-TOX**

CH_3 CH_3 C O S

CH_2—O—C—CH—CH—CH=C

CH_2 O O

5-Benzyl-3-furylmethyl-d-cis(1R,3S,E)2,2-dimethyl-3-(2-oxo,-2,2,4,5 tetrahydro thiophenylidenemethyl) cyclopropane carboxylate

TYPE: Kadethrin is a synthetic-pyrethroid compound used as a contact insecticide.

ORIGIN: Roussel Uclaf of France, 1974.

TOXICITY: LD_{50}-142 mg/kg. May cause eye and skin irritation.

FORMULATIONS: Aerosols

PHYTOTOXICITY: Not for use on plants.

USES: Outside the U.S. as a household insecticide.

IMPORTANT PESTS CONTROLLED: Flies, mosquitoes, midges, cockroaches, and other household pests.

APPLICATION: Apply as a room spray aerosol.

PRECAUTIONS: Not for sale in the U.S. Toxic to fish and bees.

ADDITIONAL INFORMATION: Provides quick knockdown. May be mixed with other nonalkaline insecticide.

NAMES

***RESMETHRIN*, BENZYFUROLINE, BIORESMETHEIN, CHRYSON, CROSSFIRE, EARTHFIRE, FMC-17370, FOR-SYN, NRDC 104, PREMGARD, PYRETHERM, RESPOND, SBP-1382, SCOURGE, SYNTHRIN**

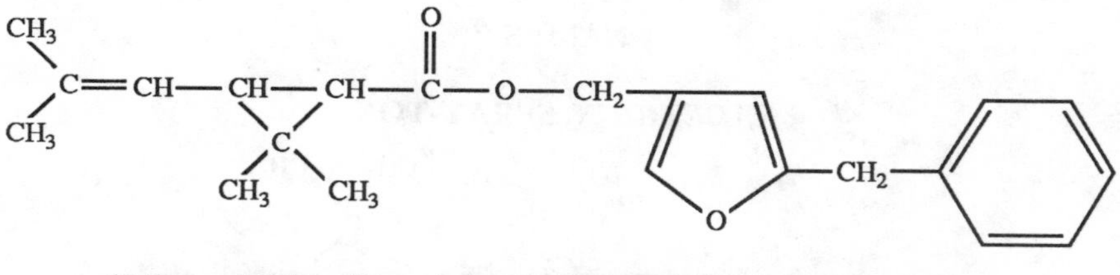

(5-Benzyl-3-furyl) methyl 2,2-dimethyl-3-(2-methylpropenyl) cyclopropanecarboxylate

TYPE: SBP-1382 is a synthetic-pyrethroid compound used as a selective, contact insecticide.

ORIGIN: National Research and Development Corporation in England, 1968. Licensed to be manufactured in the U.S. by Penick and Co., and Fairfield-American, and in Japan by Sumitomo Chemical Co.

TOXICITY: LD_{50}-2500 mg/kg.

FORMULATIONS: 2 EC, 10% WP, sometimes formulated with Allethrin.

PHYTOTOXICITY: Non-phytotoxic when used as recommended.

IMPORTANT PESTS CONTROLLED: Flies, roaches, ants, mosquitoes, wasps, gnats, spiders, centipedes, fleas, earwigs, sowbugs, moths, and others.

USES: Used as a household and garden insecticide, and in agricultural premises.

APPLICATION: Apply when insects appear, and repeat as necessary. Fill room with a mist and keep closed for 15 minutes. Used on ornamentals, both outdoors and indoors. May be used in greenhouses.

PRECAUTIONS: Do not store below 32°F. Synergists used with pyrethrins do not work well with this material. Toxic to fish.

ADDITIONAL INFORMATION: Nonirritating to the skin, eyes, or throat. Replaces the pyrethrins at the same concentration without a synergist. A contact insecticide with fast knockdown. Chemically, it is related to pyrethrin. Nonstable in sunlight, so it has limited agricultural usefulness.

RELATED MIXTURES:

1. TETRALATE — A combination insecticide being sold by Fairfield American, containing Neo-pyramin and Resmethrin, for use as a household industrial insecticide.

RELATED COMPOUNDS:

1. Chrysron-Forte-Bioresmethrin — A synthetic-pyrethroid compound sold by Sumitomo Chemical as a household and industrial insecticide.

NAMES

***TETRAMETHRIN*, BUTAMIN, DOOM, ECOTHRIN, MULTICIDE, NEO-PYNAMIN, PHTHALTHRIN, RESIDRIN, SPRIGONE, SPRITEX, TETRALATE**

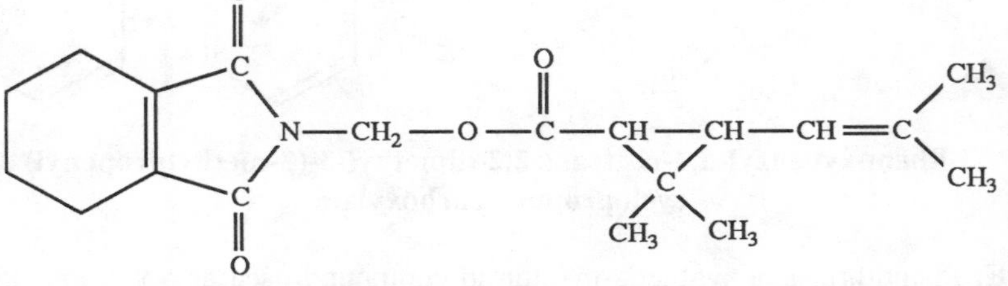

3,4,5,6-Tetra hydro-phthalimidomethyl-(1RS)-cis-trans-chrysanthemate

TYPE: Tetramethrin is an organic, synthetic-phyrethrum derivative, used as a contact insecticide.

ORIGIN: Sumitomo Chemical Co. of Japan, 1965. Licensed to be developed in the U.S. by Fairfield-American Corporation and others.

TOXICITY: LD_{50}-4640 mg/kg.

FORMULATIONS: Aerosols, sprays, and dust, 25% EC. Formulated in many combinations.

USES: Households, industrial, and agricultural premises. Also used in some countries for control of home and garden pests.

IMPORTANT PESTS CONTROLLED: Flies, wasps, mosquitoes, and garden pests.

APPLICATION: Apply when insects appear, and repeat as necessary.

PRECAUTIONS: Do not use on food or feed crops. Toxic to fish.

ADDITIONAL INFORMATION: Sold formulated with many synergists or other insecticides. Has essentially the same characteristics as allethrin or pyrethrum. Pyrethrum-like odor. Long storage ability. Processes fast knockdown ability. Promising for the control of cattle insects and stored-product pests.

NAMES

PHENOTHRIN, FENOTHRIN,
MULTICIDE CONCENTRATE F-2271,
S-2539, SUMITHRIN

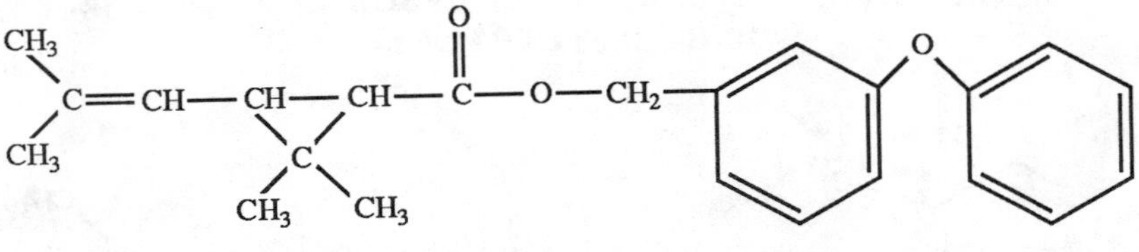

3-Phenoxybenzyl-d,1-cis,trans 2,2-dimethyl-3-(2-methylpropenyl) cyclopropane carboxylate

TYPE: Phenothrin is a synthetic-pyrethroid compound used as a fast-knockdown insecticide.

ORIGIN: Sumitomo Chemical Co. of Japan, 1971. Sold in the U.S. by Fairfield American.

TOXICITY: LD_{50}-10,000 mg/kg.

FORMULATIONS: EC, aerosols. 38% active liquid.

USES: To control household insects and for mosquito control.

IMPORTANT PESTS CONTROLLED: Most Lepidoptera and Hemiptera insects, cockroaches, bedbugs, fleas, flies, gnats, lice, mosquitoes, and others. Used as a mosquito adulticide.

RATES: Applied at 100-250 ppi ai or at .004-.061 lb a.i./A.

APPLICATION: Apply when insects appear, and repeat as necessary. For mosquitoes, apply to residential and recreational areas, swamps, marshes, waste areas, and roadsides.

PRECAUTIONS: Toxic to fish and bees. Short residual in sunlight. Do not mix with alkaline compounds.

ADDITIONAL INFORMATION: Broad spectrum. Kills both as a stomach-poison and upon contact, with quick knockdown. Sumithrin is the name of the isomer of this compound which is being developed as a public health insecticide. Sold in the U.S. as a household aerosol insect spray.

RELATED COMPOUNDS:

1. *EMPENTRIN*, VAPORTHRIN — A synthetic-pyrethroid compound developed by Sumitomo Chem. Co. of Japan to control clothes moths and related insects in indoor situations and as a moth proofer of textiles. It controls by vapor action with good residual activity.
2. GOKILAHT (cyphenothrin) — A synthetic-pyrethroid being developed by Sumitomo Chem. Co. of Japan for use as an indoor insecticide to control houseflies, cockroaches, mosquitoes, etc.
3. *PRALLETHRIN*, ETOC — A synthetic pryrethroid compound developed by Sumitomo to use as a public health insecticide against mosquitos and cockroaches.

RELATED MIXTURES:

1. PESGUARD — A combination of Sumithrin and Neopynamin, developed by Sumitomo Chemical Co. as a public health insecticide.

NAMES

ALLETHRIN, PYRETHRIN, ALLEVIATE, BIOALLETHRIN, CINERIN, CINEROLONE, ESBIOL, ESIBIOTHRIN, PALLETHRINE, PYNAMIN-FORTE, PYRESYN, PYREXCEL, PYROCIDE, SECTOL, SECTROL, SYNEROL

2-allyl-4-Hydroxy-3-methyl-2-cyclopenten-1-one ester of 2,2-dimethyl-3-(2-methyl propenyl)-cyclopropane carboxylic acid

(2-methyl-1-propenyl)-2-methoxy-4-oxo-3-(2 propenyl)-2-cyclopenten-1-yl ester or mixture of _cis_ and _trans_ isomers

TYPE: Pyrethrin and Allethrin are botanical and synthetic insecticides which have contact, stomach-poison, and fumigant action.

ORIGIN: Used in Iran around 1828. Allethrin was first synthesized in 1949. The basic manufacturers of Allethrin are Roussel Uclaf of France and Sumitomo Company of Japan. Fairfield-American, Penick, Webb Wright Corp., MGK, and Pentriss distribute in the U.S.

TOXICITY: Pyrethrin-LD_{50}-200 mg/kg. Allethrin-LD_{50}-310 mg/kg. Some people are allergic to it.

FORMULATIONS: Numerous. Often sold with piperonyl butoxide and other synergists.

PHYTOTOXICITY: Non-phytotoxic.

USES: Allethrin-Apples, beans, beets, broccoli, Brussels sprouts, cabbage, carrots, cauliflower, celery, citrus, collards, corn, crab apples, currants, endive, figs, garlic, gooseberries, grapes, guavas, horseradish, caneberries, huckleberries, kale, kohlrabi, leeks, lettuce, mangoes, mushrooms, melons, mustard greens, onions, parsley, peaches, pears, peppers, pineapples, plums, potatoes, radishes, rutabagas, spinach, stored grains (barley, corn, milo, oats, rye, sorghum, wheat), sugarbeets, sweet potatoes, tomatoes, turnips, ornamentals, and others. Pyrethrin-Bush and vine fruits, deciduous fruits and nuts, forage crops, citrus, grains, mushrooms, peanuts, tobacco, vegetables, animals, poultry, agricultural premises, ornamentals, food handling establishments, and in public health programs.

IMPORTANT PESTS CONTROLLED: Mosquitoes, flies, aphids, Mexican bean beetles, imported cabbageworms, mealybugs, beetles, thrips, fleabeetle, sod webworm, loopers, leafhoppers, lice, Colorado potato beetle, and many others.

RATES: Applied in the field at 1/2 lb actual/A and in households as an aerosol at .6 g/1000 cu ft.

APPLICATION:

1. Foliage-Apply evenly as often as necessary. There are no limitations on when it can be applied to the crop.
2. Postharvest-Apply to grain as it is being placed in storage. Also applied directly in the space over commodities in storage. Used as a seed treatment.
3. Livestock-Apply as a mist or pressurized spray at the rate of 2 oz/animal the size of an adult cow. It can be applied liberally to swine. Repeat as necessary up to twice daily. Used also as a dip treatment.
4. Premises-One of the most widely used household sprays. Surface-spray to the point of runoff or use aerosol formulations, keeping the room full of mist for 10 minutes. Used to a great extent in greenhouses.

PRECAUTIONS: Noncompatible with Bordeaux, calcium arsenate, lime, lime sulfur, and soaps. Ineffective in alkaline solutions. Very toxic to cold-blooded animals. Not compatible with lead, brass, copper, zinc, or iron. Nonstable in sunlight, so limited in agricultural usefulness.

ADDITIONAL INFORMATION: Imported from Africa and South America into the United States, making it relatively expensive. Comes from the flowers of pyrethrium plants. Effective on most insects it comes in contact with, however, roaches seem somewhat resistant. Due to its relatively high cost, the use of pyrethrins alone has been restricted. Rapid knockdown effects are obtained. A synergist may be added to enhance this knockdown. Rapidly broken down by sunlight so there are no residue problems. Paralyzes insects before killing them.

Allethrin is as toxic to insects as natural pyrethrins, but it has longer residual effects. Allethrin is the allyl homolog of Cinerin I which is a synthetic analog of one of the four toxic constituents of pyrethrin.

RELATED MIXTURES:

1. PYRENONE — An insecticide sold by Fairfield-American, which contains piperonyl butoxide and pyrethrins in ratios ranging from 5:1 to 20:1 by weight. Pyrenone crop spray, for example, contains 6% pyrethrins and 60% piperonyl butoxide and is registered for several uses on vegetables, stored grains, and ornamentals. Other Pyrenone products hold registrations on a wide variety of fruits and vegetables, for control of certain livestock pests, for postharvest uses in stored grain and food products, and for the control of household insects. Registered to tank mix with residual insecticides at a reduced rate to flush the insects out of hiding, giving better control.

RELATED COMPOUNDS:

1. Pynamin-Forte — A synthetic-pyrethroid compound made by Sumitomo Chemical Co. in Japan for use as a household and industrial insecticide.
2. BUTACIDE (pyeronyl butoxide).— Produced by Fairfield American, this product is registered to tank mix with synthetic-pyrethroid insecticides to act as a synergist that interrupts the natural defense mechanism of the insects.

NAMES

***ROTENONE*, BARBASCO, CUBE ROOT, CUBOR, DERRIN, DERRIS, EXTRAX, FISH-TOX, HAIARI, MEXIDE, NOXFIRE, NOXFISH, PRO-NOX FISH, RO-KO, ROTEFIVE, ROTESSENOL**

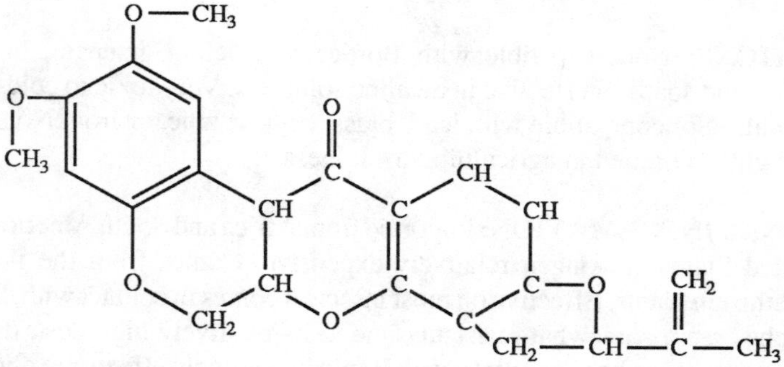

1,2,12,12a,Tetrahydro-2-isopropenyl-8,9-dimethoxy-(1) benzopyrano-(3,4,-6)furo(2,3-6) (1)benzopyran-6(6aH) one

TYPE: Rotenone is a botanical insecticide having both contact and stomach-poison activity.

ORIGIN: First used on crops in British Malaya in 1848. England patented it in 1911. The chemical nature was determined in 1932. Sold in the U.S. by Fairfield America, Prentiss Drug, Penick and Co., and others. Supplied by Foreign-Domestic Chemicals, Inc.

TOXICITY: LD_{50}-132 mg/kg. Very toxic to fish. Swine are highly susceptible.

FORMULATIONS: Dusts 1/2-1%, 4-5% WP.

PHYTOTOXICITY: Non-phytotoxic.

USES: Bush and vine crops, citrus, deciduous fruits, forage crops, mushrooms, asparagus, beans, beets, corn, eggplant, mustard, peas, potatoes, radishes, strawberries, tomatoes, and other vegetables. Also used to control undesirable fish.

IMPORTANT PESTS CONTROLLED: Ants, Colorado potato beetle, beetles, weevils, slugs, Japanese beetles, fleabeetles, loopers, cabbageworm, mosquitoes, thrips, fleas, lice, flies, and many others. Also used to control undesirable fish species and eradicate them from lakes, streams, and reservoirs.

RATES: Applied at .5-2 lb actual/A.

APPLICATION:

1. Foliage-Apply evenly at a uniform rate. Repeat as necessary.
2. Premises-Apply as a space or surface spray. Used also in fogging machines for mosquito control, or applied to pond water as a mosquito larvicide. Used on premises to control fire ants.
3. Dogs and cats-Apply as a dip, spray, dust, or by backrubbers.

PRECAUTIONS: Incompatible with lime. Long used as a fish poison. May be used only by permit to control fish. Toxic to pigs.

ADDITIONAL INFORMATION: Found in 68 species of legume plants in their root systems. Most commercial supplies come from Derris and Cube root as well as tembo. Cube root is used mostly in the U.S. Acts somewhat as a repellent and as an acaricide. Short residue (no longer that 1 week), since it is broken down in the presence of oxygen and sunlight. Kills insects by inhibiting the utilization of oxygen by the body cells, or by depriving the tissues of oxygen. Slow acting. Breaks down in storage. Ineffective if applied in an alkaline solution. Selectively used to control trash fish from game fish. Insects rarely become immune to this material. Non-toxic to bees. Used in fish ponds to kill undesirable fish species 2-4 weeks prior to restocking.

RELATED COMPOUNDS:

1. RYANIA, RYANODINE, RYAN — A neutral alkaloid isolated from the plant Ryania speciosa native to South America. It is a slow acting stomach poison insecticide. Registered by Dunhill Chemical Co. on citrus, apples, pears, walnuts, and corn to control thrips, codling moth and European corn borer.
2. SABADILLA, VERATRAND — A botanical alkaloid derived from a lily seed in South American. Registered by Dunhill Chemical Co. on citrus to control thrips.
3. NICOTINE-A Botanical insecticide derived from tobacco, used extensively a number of years ago. Due to its toxicity, it has largely been replaced by newer insecticides. Produced by Black Leaf Products Co. as Black Leaf 40. Mostly used in home gardens. Used to a limited extent in greenhouses and in veterinary compounds, and in home and garden preparations. LD_{50}-55 mg/kg. Registered on apples, apricots, asparagus, beans, cabbage, melons, cherries, citrus, corn, cranberries, cucumbers, currants, eggplant, gooseberries, grapes, onion, peas, peaches, pears, peppers, plums, prunes, spinach, squash, strawberries, tomatoes, turnips, poultry, and ornamentals.

NAMES

ACID LEAD ARSENATE, ARSENATE OF LEAD, ARSINETTE, BASIC LEAD ARSENATE, GYPSINE, LEAD ARSENATE, SOPRABEL, STANDARD LEAD ARSENATE

I. $PhHAs0_4$-Acid lead arsenate
II. $Pb_4(PbOH)(AsO_4)_3{}'H_2O$ (Basic lead arsenic)

A mixture of the two are Basic Lead Arsenate containing at least 14% arsenic.

TYPE: Lead arsenate is an inorganic arsenical used as a stomach-poison insecticide.

ORIGIN: First used about 1894.

TOXICITY: LD_{50}-100 mg/kg. .1 gram is lethal to man. May cause skin irritation.

FORMULATIONS: Pastes containing 14% arsenic and higher, and powders or dusts containing 32% arsenic or higher.

PHYTOTOXICITY:

1. Acid form-Non-phytotoxic when used properly, except in coastal areas on tender foliage. Injury has been reported on beans.
2. Basic form-Safer on the foliage than the acid form, especially in coastal areas.

USES: No longer used in the U.S. Used in other countries on apples, apricots, asparagus, avocadoes, blueberries, celery, cherries, cranberries, currants, eggplant, gooseberries, grapes, loganberries, mangoes, nectarines, peaches, pears, peppers, plums, prunes, quinces, raspberries, strawberries, tomatoes, tobacco, turf, and ornamentals.

IMPORTANT PESTS CONTROLLED: Codling moths, Japanese beetles, curculios, cankerworms, leaf rollers, hornworms, potato beetles, tomato fruitworms, budworms, scale, weevils, grasshoppers, fruit flies, and many others.

RATES: Applied at 3-60 lb actual/A or 1-60 lb actual/100 gal of water.

APPLICATION: Apply at 10-14 day intervals as necessary for control. Cover thoroughly. May be sprayed on the soil for soil-borne insects.

PRECAUTIONS: It may need a safener, such as chemically hydrated lime added, when applied to tender foliage of certain species. Do not feed treated crop residue to livestock. Any form of arsenic used on tobacco should be used with caution because of the possibility of leaving an objectionable arsenic residue on the leaves. Readily becomes fixed in the soil. Extremely toxic to wildlife and bees. Not for sale in the U.S.

ADDITIONAL INFORMATION: Most effective on chewing insects, since it is a stomach-poison with very little contact activity. The acid form is used primarily because it reacts more readily and is more insecticidal than the basic form. Thorough applications have grub-proofed turf for periods of 3-4 years. Remove all residue from edible parts by washing, brushing, etc. The basic form is more water soluble than acid. It increases the effects of wettable sulfur when applied with it. Generally considered the least phytotoxic of the arsenicals. Noncorrosive. Compatible with most other pesticides.

NAMES

CRYOLITE, KRYOCIDE, Na_3 ALF_6, PROKIL

Sodium aluminofluoride or sodium fluoaluminate

TYPE: Cryolite is an inorganic stomach and contact-poison insecticide-acaricide.

ORIGIN: First used in 1959. Pennwalt Corp. is the main producer.

TOXICITY: LD_{50}-10,000 mg/kg.

FORMULATION: WP and dusts.

PHYTOTOXICITY: Some injury has been reported on peaches, grapes, and apples. Burning of corn has been reported in damp climates.

USES: Apples, beans, broccoli, cabbage, cantaloupes, cauliflower, citrus, collards, cranberries, cucumbers, grapes, lettuce, melons, mustard greens, peaches, pears, peppers, radishes, squash, strawberries, tomatoes, turnips, trees, ornamentals, and watermelons.

IMPORTANT PESTS CONTROLLED: Codling moths, orange tortrix, beetles, weevils, Mexican bean beetles, mites, thrips, corn earworms, cabbageworms, cabbage looper, leaf rollers, plum curculio, armyworms, flea beetles, and hornworms.

RATES: Applied at 5-79 lbs/A.

APPLICATION: Use as a dust as soon as insects first appear, repeating as necessary. Cover thoroughly.

PRECAUTIONS: Some compatibility problems have arisen, although it is the most compatible of the fluorine compounds. Do not apply to leafy vegetables after the edible parts have started to form. Remove excess residues by washing, brushing, etc. Incompatible with alkaline compounds. Toxic to bees. Do not use in combination with lime or compounds containing free lime.

ADDITIONAL INFORMATION: A naturally occurring compound mined in Greenland. Manufactured synthetically in the U.S. and other countries. Most effective on chewing insects.

NAMES

DRI-DIE, SANTOCEL C, SG-67, SILICA AEROGEL, SILIKIL, SPROTIVE DUST SG-67

(Contains a mixture of silica aerogel and ammonium fluosilicate.)

TYPE: Silica Aerogel is an inorganic insecticide, killing upon contact.

ORIGIN: Produced by a number of manufacturers, 1956.

TOXICITY: LD_{50}-3160 mg/kg.

FORMULATIONS: 100% dusts and aerosols. Sometimes mixed with other pesticides which act as a knockdown agent.

PHYTOTOXICITY: Mushrooms have been injured. Not used on plants, except experimentally, although no undue injury has been reported.

USES: Agricultural premises, cattle, goats, sheep, and hogs.

IMPORTANT PESTS CONTROLLED: Cockroaches, grasshoppers, scorpions, earwigs, mites, snails, screwworms, spiders, termites, wasps, bedbugs, ants, chicken mites, fleas, flies, silverfish, and others.

RATES: Applied at 1.25 lb/1000 sq ft at 40 lb/A or 4 lb/100 gal of water.

APPLICATION: Structural pest control-Apply where insects will walk through the material. Monthly treatments are usually necessary to eliminate infestation.

PRECAUTIONS: Termites already in the wood will not be controlled since they won't walk on the material.

ADDITIONAL INFORMATION: Prepared by treating sodium silicate with sulfuric acid, and then drying and grinding the residual material. Acts on insects by removing the oily, protective film covering their bodies which normally prevents the loss of water. Kills the insects by dessication. It surpasses common insecticides on roaches. Insects cannot develop resistance to it. It is often made coated with fluorine (ammonia fluosilicate) to give the material insecticidal properties. Silica alone doesn't kill insects chemically, but is a physical means of control. Roaches die within 3-4 hours of crawling over the material. Silica (sand) alone, has been used for many centuries as a means of insect control. Poor results have been obtained when used on sucking insects, such as aphids, since they are able to replenish their body moisture. For use on livestock, it is formulated with other ingredients and applied as a screwworm or ear tick remedy.

RELATED COMPOUNDS:

1. BORAX OR SODIUM TETRABORATE DECAHYDRATE OR BORIC ACID — Used as a powder or dust formulation for cockroaches, ants, and silver fish control in indoor areas. Applied into cracks and crevices. Applied in poultry houses to control maggots in manure and in other areas where maggots may appear. Used in dog kennels to aid in the control of dog hookworm larvae. Applied as a dip to freshly sawn lumber to prevent beetle infestation. Slow acting requiring 7-10 days for complete control, but it will remain active for an extended period of time. Being tested for termite control.
2. SODIUM FLUORIDE — An older insecticide used on agricultural premises, buildings, and for indoor pest control of ants, cockroaches, termites, silverfish, crickets, sowbugs, and similar insects.
3. SODIUM FLUOSILICATE — An older insecticide used on ornamentals and buildings for the control of earwigs, cutworms, sowbugs, ants, and cockroaches.
4. SULFUR — Widely used as a fungicide, but also controls or prevents buildup of mite populations.

DIPHENYL COMPOUNDS and other Nonphosphate Insecticides

NAME

DDT

Cl—C_6H_4—CH(—C_6H_4—Cl)—CCl_3

Dichlorodiphenylthrichloroethane

TYPE: DDT was the most widely used chlorinated-hydrocarbon insecticide, expressing both stomach-poison and contact activity, as well as long residual effects.

ORIGIN: Geigy Chemical Company, 1940. Widely used by 1944. Still produced in a number of countries.

TOXICITY: LD_{50}-113 mg/kg.

FORMULATIONS: Numerous formulations are available.

PHYTOTOXICITY: Injury has been reported on cucurbits, young tomatoes, and beans.

USES: No longer used in the U.S., except for emergency public health uses. Used on almost all crops in other parts of the world.

IMPORTANT PESTS CONTROLLED: Peach tree borers, codling moths, flea-beetles, leafhoppers, psyllid, corn earworms, corn borers, thrips, flies, mosquitoes, leaf miners, Japanese beetles, spittlebugs, chinch bugs, bollworms, lygus bugs, and many others.

RATES: Applied at 1-2 lbs a.i./A.

APPLICATION: Apply uniformly when insects first appear. Repeat as often as necessary. Also used as a soil treatment and a seed treatment.

PRECAUTIONS: May accumulate in the soil. Incompatible with alkaline materials. Do not store in iron containers. Toxic to bees. Accumulates in the body fat of animals. Not for sale or use in the U.S.

ADDITIONAL INFORMATION: One of the first new organic insecticides put on the market. At one time it was one of the most widely used insecticides in the world. Little toxicity to most Orthoptera (grasshoppers, crickets, cockroaches), boll weevils, Mexican bean beetles, and most aphids. Stable under most conditions. Acts on the central nervous systems of the insects. Many insects are DDT resistant due to prolonged use of the insecticide. Used in other countries as a human-body insecticide.

NAMES

METHOXYCHLOR, DMDT, MARLATE, METHOXCIDE, METHOXO, MOXIE

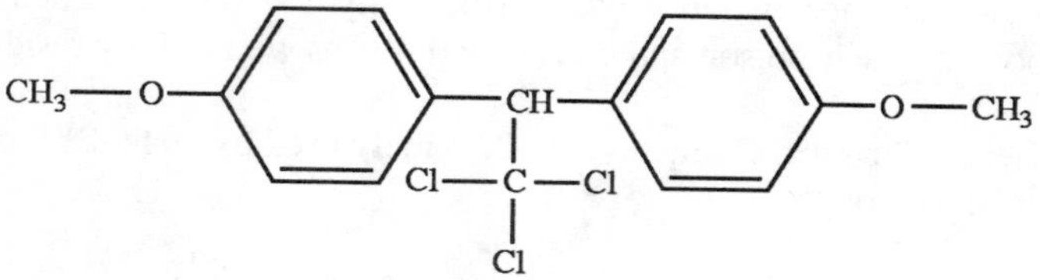

CH_3—O—C6H4—CH(CCl3)—C6H4—O—CH_3

2, 2-bis(p-Methoxyphenyl)-1,1,1-trichloroethane

TYPE: Methoxychlor is a chlorinated-hydrocarbon insecticide having long residual activity.

ORIGIN: CIBA-Geigy Chemical Company and E. I. DuPont de Nemours and Co., 1944. Kincaid Enterprises is the principal producer today.

TOXICITY: LD_{50}-6000 mg/kg. The fatal dose for man is estimated to be about 1 lb if injected at one time.

FORMULATIONS: 25% WP, 50% WP, 2 EC, 3-5% dusts, 2 and 3% aerosols.

PHYTOTOXICITY: Considered to be Non-phytotoxic.

USES: Alfalfa, apples, apricots, asparagus, barley, beans, beets, blackberries, black-eyed peas, blueberries, boysenberries, broccoli, Brussels sprouts, cabbage, cantaloupes, carrots, cauliflower, cherries, clover, collards, corn, cowpeas, cranberries, cucumbers, currants, eggplants, gooseberries, grapes, grasses, huckleberries, kale, kohlrabi, lettuce, lima beans, loganberries, melons, nectarines, oats, peaches, peanuts, pears, peas, peppers, pineapples, plums, potatoes, prunes, pumpkins, quince, radishes, raspberries, rice, rutabagas, rye, sorghum, soybeans, spinach, squash, strawberries, sweet potatoes, tomatoes, turnips, velvet beans, wheat, yams, beef and dairy cattle, sheep, goats, swine, greenhouses, mushroom houses, outdoor fogging, grain bins, and agricultural premises.

IMPORTANT PESTS CONTROLLED: Houseflies, lice, ticks, weevils, stored-grain beetles, leafhoppers, Japanese beetles, armyworms, codling moths, plum curculio, spittlebugs, scale (crawlers), lygus bugs, and many others.

RATES: Applied at 1/4-1/2 lb actual/A.

APPLICATIONS: Apply to crops with common application equipment. Start application at first signs of infestation and repeat in 7-14 days intervals as needed. It can be sprayed around livestock premises or directly upon livestock for control of external parasites. Used also in grain storage bins and in household insect sprays. Used as a seed treatment on a number of grain crops.

PRECAUTIONS: Do not use in dipping vats. Do not add grain to a treated bin for at least 24 hours, or until the walls have dried out thoroughly. Avoid contamination of fish-bearing waters. Do not use with any product that is incompatible with oil. Do not apply within 14 days of sulfur or sulfur product applications. Aphids and mites are not controlled. Toxic to fish and bees.

ADDITIONAL INFORMATION: Closely related to DDT, but only 1/25 to 1/50 as toxic to mammals, and accumulates less in the body fat. Compatible with other insecticides and fungicides, except strong alkaline solutions.

RELATED MIXTURES:

1. DYMET — A combination of methoxychlor and diazinon developed by Mallenckrodt, Inc. for use on ornamentals.

NAMES

DICOFOL, ACARIN, CARBAX, HILFOL, KELTHANE, MITIGAN

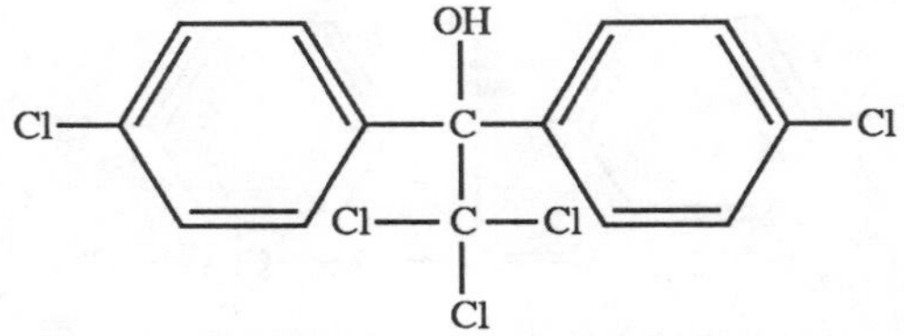

OH
Cl— —C— —Cl
Cl—C—Cl
Cl

1,1-bis(p-Chlorophenyl)-2,2,2-trichloroethanol

TYPE: Kelthane is a chlorinated-hydrocarbon acaricide, controlling upon contact.

ORIGIN: Rohm and Haas Company, 1957.

TOXICITY: LD_{50}-684 mg/kg.

FORMULATION: 18.5%WP, 35% WP, 18.5 and 42% EC, 30% dust.

PHYTOTOXICITY: Injury has occurred on eggplant and pears. Do not use on canaeti juniper or Chinese holly.

USES: Alfalfa, apples, apricots, beans, caneberries, chestnuts, citrus, clover, cotton, crab apples, cucumbers, dewberries, eggplant, figs, filberts, grapes, hickory nuts, hops, melons, mint, nectarines, peaches, pears, peanuts, pecans, peppers, plums, prunes, pumpkin, quince, squash, strawberries, tea, tomatoes, walnuts, watermelons, turf, and ornamentals.

IMPORTANT PESTS CONTROLLED: Mites. No insecticidal activity.

RATES: Applied at .5-2 actual/100 gal of water or .6-8 lb actual/A.

APPLICATION: Apply thoroughly, covering the undersides of leaves. Apply when mites first appear in threatening numbers, and repeat as often as necessary. May be applied by air.

PRECAUTIONS: Do not feed crop residues to dairy or slaughter animals. Do not mix with lime. Toxic to fish. Usage in the U.S. is under investigation at this time.

ADDITIONAL INFORMATION: Gives good initial kill, and long residual action against mites. Compatible with other commonly used pesticides. Does not harm beneficial insects. Nonsystemic.

NAMES

CHLOROBENZILATE, ACARABEN, AKAR, BENZILAN, BENZ-O-CHLOR, FOLBEX, KOPMITE

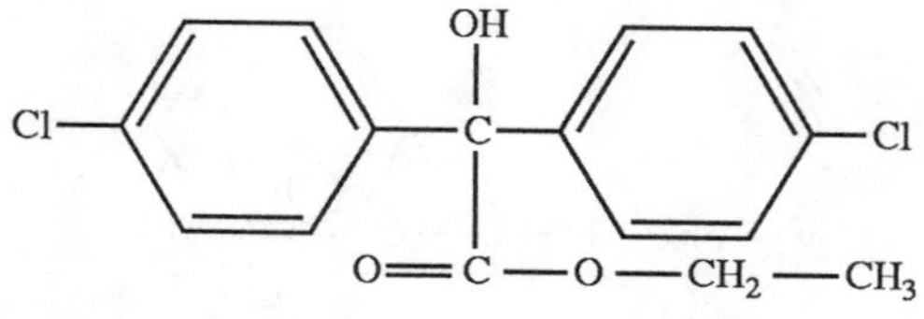

OH

Cl— —C— —Cl

O=C—O—CH_2—CH_3

Ethyl 4,4'-dichlorobenzilate

TYPE: Chlorobenzilate is a chlorinated-hydrocarbon acaricide, expressing contact activity.

ORIGIN: CIBA-Geigy Corp., 1952.

TOXICITY: LD_{50}-700 mg/kg.

FORMULATIONS: 4 EC, 25% EC.

PHYTOTOXICITY: Injury has been noted on peaches, plums, and prunes. Do not use on the McIntosh, Delicious, or Jonathan varieties of apples. Russeting of pears has been noted. Injury to roses has occurred in the western states. Do not apply to Kapareil variety of almonds.

USES: Almonds, apples, citrus, cherries, cotton, melons, pears, walnuts, turf, ornamentals, and premises outside the U.S. Inside the U.S. the only remaining use is in citrus.

IMPORTANT PESTS CONTROLLED: Most species of mites.

RATES: Applied at 1/4-1/2 lb actual/100 gal of water or 1/2-5 lb actual/A.

APPLICATION: Apply with common application equipment at a uniform rate. Apply when mites first appear, and repeat as necessary. Apply in sufficient water to insure thorough coverage.

PRECAUTIONS: Compatible with other insecticides and fungicides, except the highly alkaline materials. Hydrated lime slightly retards acaricidal activity. Do not apply to ornamentals when temperatures exceed 90°F.

ADDITIONAL INFORMATION: May be applied to some crops up to the day of harvest. Safe to beneficial insects. Formulated on strips to be used in beehives for control of bee disease. Controls all stages of mites, including their eggs. Used to prevent clover mites from migrating to adjoining buildings.

NAMES

***BROMOPROPYLATE*, ACAROL, FOLBEX, NEORON**

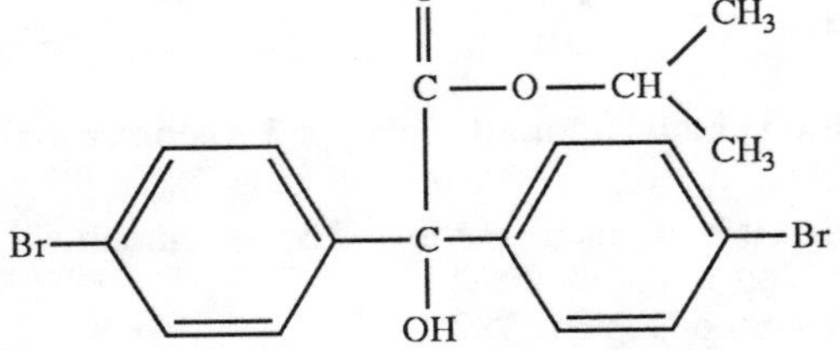

Isopropyl 4,4'-dibromobenzilate

TYPE: ACAROL is a chlorinated-hydrocarbon being used as a long residual, contact acaricide.

ORIGIN: CIBA-Geigy Corp., 1968.

TOXICITY: LD_{50}-5000 mg/kg.

FORMULATIONS: 2 EC, 50% EC.

PHYTOTOXICITY: Slight injury has been observed on certain varieties of apples, plums, and ornamentals.

USES: Used on numerous crops outside the U.S., such as fruit crops, citrus, hops, cotton, beans, cucurbits, tomatoes, strawberries, and ornamentals.

IMPORTANT PESTS CONTROLLED: Most species of mites.

RATES: Applied at 1/2-1 lb a.i./A or at 1/4-3/4 lb/100 gal of water.

APPLICATION: Apply when mites appear and repeat as necessary.

PRECAUTIONS: Not for sale or use in the U.S. Do not graze treated areas.

ADDITIONAL INFORMATION: Considerable residual activity. Has longer residual activity and improved crop tolerance over chlorobenzilate and chloropropylate. Aerial applications have proven effective. Store at temperatures above 32°F. May be mixed with other pesticides.

NAMES

ETHOPROXYFEN, MTI-500, TREBON

CH_3

H_3— CH_2—O— —C— CH_2—O—CH_2— —O—

CH_3

2-(4-Ethoxyphenyl)-2-methylpropyl 3-phenoxybenzyl ether

TYPE: Ethoproxyfen is a stomach-poison and contact insecticide.

ORIGIN: Mitsui Toatsu of Japan, 1982.

TOXICITY: LD_{50}-40,000 mg/kg.

FORMULATIONS: .5% dust, 1.5% granule, 10-30% EC, 10-30% WP.

PHYTOTOXICITY: Non-phytotoxic when used as directed.

USES: Rice outside the U.S. Experimentally being used on cotton, vegetables, fruit crops, corn, soybeans, tobacco, potatoes, and others. Also used on livestock for tick, flea and cockroach control.

IMPORTANT PESTS CONTROLLED: Bollworm, budworm, cabbage looper, aphids, lygus, whitefly, boll weevil, armyworms, cutworms, codling moth, oriental fruit moth, leafminer, psylla, leafhoppers, Colorado potato beetle, rice water weevil, rice stem maggot, and others.

RATES: Applied at .075-.2 kg/ha.

APPLICATIONS: Used as a foliar spray. Repeat as necessary.

PRECAUTIONS: Not for sale or use in the U.S. Somewhat toxic to fish.

ADDITIONAL INFORMATION: Excellent residual activity. Stable in alkaline solutions.

NAMES

***TETRADIFON*, CHILDION, DOUBLE, DUPHAR, MIXAN, MURFITE, TEDION**

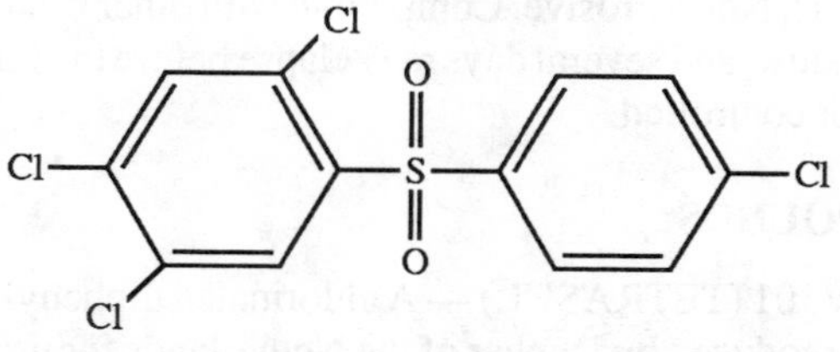

p-Chlorophenyl 2,4,5-trichlorophenyl sulphone

TYPE: Tedion is a chlorinated-hydrocarbon acaricide, effective on all forms of mites, except adults.

ORIGIN: N. V. Philips Duphar of the Netherlands, 1954. Licensed to be sold in the U.S. by Uniroyal Chem. Co.

TOXICITY: LD_{50}-14,700 mg/kg.

FORMULATIONS: 50% WP, 1 EC, 20% WP.

PHYTOTOXICITY: A few varieties of white roses have been slightly injured; otherwise, it is completely safe on crop plants. Do not spray open chrysanthemum blossoms or to Cistus, Dahlia, Ficus, Kalanchoe or Primulas.

USES: Apples, apricots, caneberries, cherries, citrus, crab apples, cucumbers, figs, gooseberries, grapes, hops, melons, mint, nectarines, peaches, pears, plums, prunes, pumpkins, quince, squash, strawberries, tomatoes, and ornamentals. Used outside the U.S. on these and other crops.

IMPORTANT PESTS CONTROLLED: Most species of mites. Laboratory observation has shown that treated adult female mites deposit eggs that either fail to develop, or are delayed in their normal hatching period. Effective on all stages of mites, except the adults.

RATES: Applied at 1/8 to 1 lb actual/100 gal of water.

APPLICATION: Applied in a preventative program when mite eggs are present and hatching, and when motile forms, particularly adults, average less than 1-2 per leaf. Apply in sufficient water to give thorough coverage.

PRECAUTIONS: Do not apply to citrus more than once per fruit season. Do not apply undiluted spray to plants. Do not graze treated areas.

ADDITIONAL INFORMATION: Long residual effects, although it is not fully systemic. Not effective against over-wintering eggs. Known to give mite control from six months to one year on citrus, and thirty to sixty days on deciduous fruits. Noncorrosive. Compatible with other pesticides. Kill of adult mites may be slow, and several days may elapse before the full effect is realized. Insects are not controlled.

RELATED COMPOUNDS:

1. ANIMERT-V101 (TETRASUL) — A chlorinated diphenyl compound related to tedion and produced by Duphar of the Netherlands for use as a mite ovicide. Available as a 20 WP or a 20% EC.

NAMES

CHLORFENSON, CPCBS, OVEX, OVOTRAN, SAPPIRAN

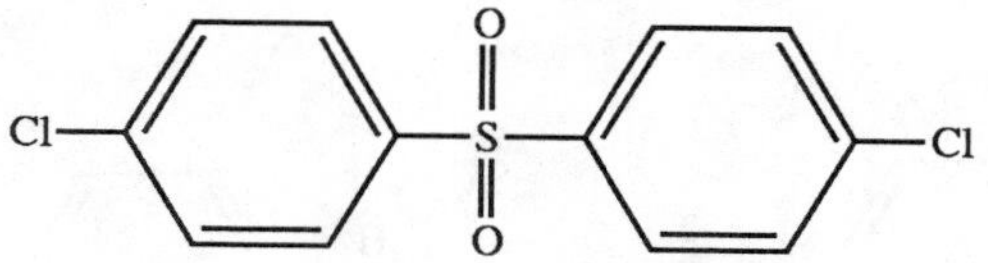

Cl— —S(=O)(=O)— —Cl

p-Chlorophenyl-p-chlorobenzene sulfonate

TYPE: Sappiran is a chlorinated-diphenyl compound used as a contact acaricide.

ORIGIN: Nippon Soda Co., Ltd. of Japan, 1969. No longer sold in the U.S.

TOXICITY: LD_{50}-8800 mg/kg.

FORMULATION: 50% WP.

PHYTOTOXICITY: Russetting of apples and pears, and injury to hops has been reported. Leaf-drop has been observed on roses.

IMPORTANT PESTS CONTROLLED: Mites.

USES: In Japan on citrus, apples, pears, peaches, grapes, strawberries, vegetables, and ornamentals.

APPLICATION: Apply when mites appear, and repeat as necessary.

PRECAUTIONS: Not for sale or use in the U.S. Toxic to fish.

ADDITIONAL INFORMATION: Effective on mite eggs. Not effective as an insecticide, so bees are not harmed. Can be mixed with alkaline chemicals, such as Bordeaux and lime sulfur. Long residual effectiveness.

RELATED MIXTURES:

1. NEOSAPPIRAN-A combination of CPCBS and DCPM, sold by Nippon Soda, Ltd. in Japan, for the control of mites on a number of fruit crops.

NAMES

DIFLUBENZURON, DIFLURON, DIMILIN, PH 60-40, LARVAKIL, MICROMITE, VIGILANTE

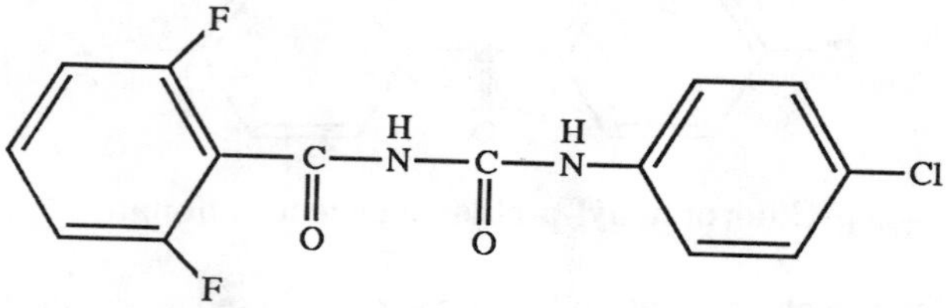

N(((4-Chlorophenyl)amino)carbonyl)-2,6-difluorobenzamide

TYPE: Benzoylurea-type insecticide interfering with chitin deposition.

ORIGIN: DUPHAR B.V., The Netherlands; 1972. research discovery. Marketed in the USA by Uniroyal Chemical Co. for crop uses, and by American Cyanamid for livestock application (cattle bolus).

TOXICITY: LD_{50}-4640 mg/kg bw. Not mutagenic ot teratogenic.

FORMULATIONS: WP25%, ODC45%, 2F, SC48%,bolus.

PHYTOTOXICITY: Non-phytotoxic at the recommended rates and uses.

USES: Larvicide in forestry, on pastures, in mushroom houses, in cotton, soybeans and chrysanthemums. Used against mosquito and fly larvae in non-crop areas. Being tested experimentally in citrus, pome fruits, walnuts, grasshoppers and on household pests. Used outside the U.S. on these and a number of additional crops.

IMPORTANT INSECTS CONTROLLED: Gypsy moth, boll weevils, army worms, leafworms, soybean caterpillar complex, cabbage caterpillars, Mexican bean beetle, mosquitoes, flies and many other insects. Being tested on rust mites, codling moth, grasshoppers, fleas, cockroaches and lice.

RATES: Applied at 0.02-0.125 lb a.i./A.

APPLICATION: Apply around oviposition time of adults for ovicidal activity or at early larval instar stages for larvicidal activity. Thorough coverage is necessary.

PRECAUTIONS: No effects on adult insects. Toxic to crustaceans.

ADDITIONAL INFORMATION: This product interferes with the formation of the insect's cuticle. Active on the larval stages of development, causing an inability to moult successfully. Does not enter the plant, so sucking insects are not controlled. In some cases also shows clear ovicidal activity, either directly on the eggs or by action through the female. Feeding will continue for a short time after application (until the next moult), so results may not be visible immediately. Has a long residual activity. Relatively harmless to beneficial insects. Relatively non-toxic to wildlife. Can be incorporated in integrated pest management systems.

NAMES

***FENAZOX*, FENTOXAN**

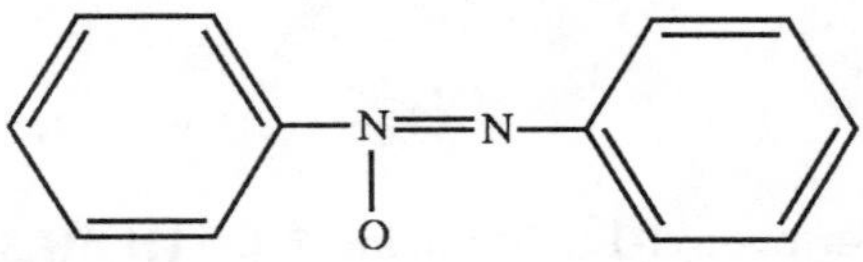

Azobenzene

TYPE: Fenazox is a diphenyl compound used as contact insecticide/acaricide.

ORIGIN: Fahberg-List of Germany, 1976.

TOXICITY: LD_{50}-885 mg/kg.

FORMULATION: 40% EC.

PHYTOTOXICITY: Non-phytotoxic when used as directed.

USES: Outside the U.S. on fruit trees, vegetables, ornamentals, soybeans, and hops.

IMPORTANT INSECTS CONTROLLED: Mites and whiteflies.

RATES: Applied at .2-.6% concentration.

APPLICATION: May be used both in the greenhouse and in the field. Apply when pests appear and repeat on a 5-7 day schedule.

PRECAUTIONS: Not for sale or use in the U.S. Toxic to fish.

ADDITIONAL INFORMATION: Non-toxic to bees. Nonsystemic. Good ovicidal activity against mites while adult mites are not completely controlled. May be mixed with other pesticides.

NAMES:

***TRIC*, WOODLUCK**

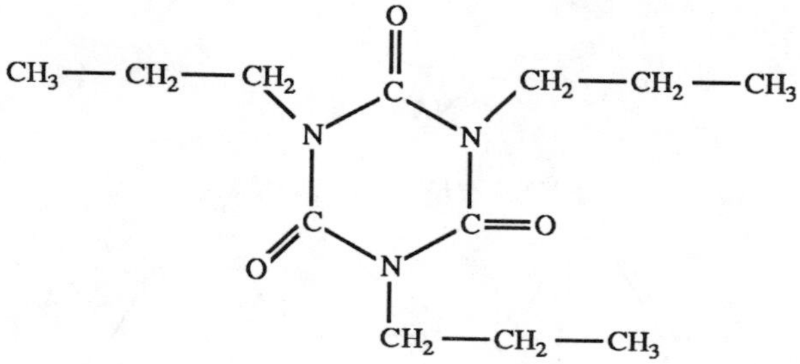

$$CH_3-CH_2-CH_2-N \quad \text{(1,3,5-tripropyl-1,3,5-triazine-2,4,6-trione ring: } N(CH_2CH_2CH_3)-C(=O)-N(CH_2CH_2CH_3)-C(=O)-N(CH_2CH_2CH_3)-C(=O)\text{)}$$

1,3,5-Tri-n-propyl-1,3,5-triazine-2,4,6(1H,3H,5H)-trione

TYPE: TRIC is a triazine compound used as a termaticide.

ORIGIN: Chugai Pharmaceutical of Japan 1984.

TOXICITY: LD_{50} 2332 mg/kg.

FORMULATION: 4% oil solution, 4% EC.

IMPORTANT INSECTS CONTROLLED: Termites and wood rot fungi.

USES: Used in Japan to treat wood for termite control. Experimentally being tested as a soil treatment for termites.

APPLICATION: Used as a dip or painted on the wood.

PRECAUTIONS: Not for sale in the U.S.

ADDITIONAL INFORMATION: Treated wood strongly repels termites. Termites are immediately killed when they come in contact with the material. High resistance to weathering. Long lasting control.

NAMES

CHLORFENETHOL, QIKRON

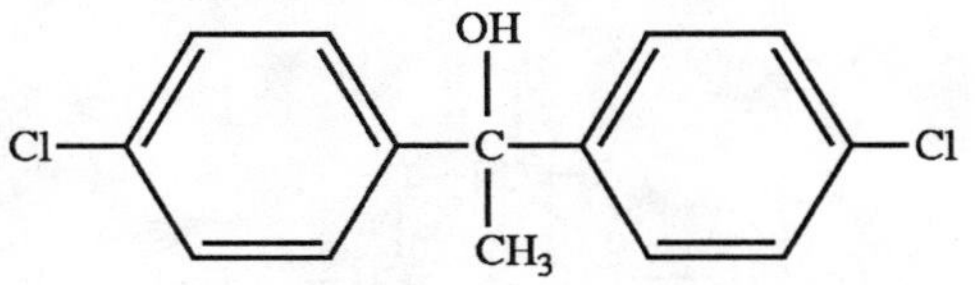

1,1-bis-(4-Chlorophenyl) ethanol

TYPE: Qikron is a chlorinated-diphenyl compound used as a contact acaricide.

ORIGIN: Nippon Soda of JAPAN, 1969.

TOXICITY: LD_{50}-958 mg/kg.

FORMULATION: 25% WP.

IMPORTANT PESTS CONTROLLED: Mites.

USES: In Japan on fruit trees.

APPLICATION: Apply when mites appear, and repeat as necessary.

PRECAUTIONS: Not for sale in the U.S. Toxic to fish.

ADDITIONAL INFORMATION: Can be tank mixed with other pesticides. Broad spectrum of mite control. Effective on adult mites.

RELATED MIXTURES:

1. MITRAN — A combination of QIKRON and SAPPIRAN developed by NIPPON SODA for use in Japan to control mites.

NAMES

METALDEHYDE, ANTIMILACE, ARIOTOX, BULLETS, CEKUMETA, DEADLINE, HALIZAN, METASON, NAMEKIL

```
                CH3
                 |
            O—CH—O
            |        |
CH3—CH        CH—CH3
            |        |
            O—CH—O
                 |
                CH3
```

Metaldehyde or Metacetaldehyde

TYPE: Metaldehyde is a stomach-poison, effective on snails and slugs.

ORIGIN: Europe, 1940. Produced by a number of companies today.

TOXICITY: LD_{50}-1000 mg/kg.

FORMULATIONS: Baits 1-20%, 1-10% solutions, 1-20% dusts, 4% foam. 4% granules.

PHYTOTOXICITY: Formulated with other pesticides. Considered non-phytotoxic at the recommended rates. The flowers of certain orchids may be injured.

USES: Used on all vegetables, bananas, avocadoes, artichokes, apples, cherries, turf, caneberries, citrus, dewberries, strawberries, and ornamentals.

IMPORTANT PESTS CONTROLLED: Slugs and snails. Very slight insecticidal properties. Also used in certain areas to control frogs, leeches, and fish.

RATES: Use the 4% bait at 40 lb/Acre.

APPLICATION: Apply in the late afternoon or at night. Good sanitation will add to the effectiveness of the treatment. Do not bring into direct contact with the foliage.

PRECAUTIONS: Do not apply just before rain. Do not use bait in an area where it may be eaten by animals or children. This material is combustible.

ADDITIONAL INFORMATION: Considered both an attractant and a toxicant to slugs and snails. Warm temperatures give the best control. The baits are most effective under arid conditions. The dusts are less effective on snails. Non-toxic to fish.

NAMES

AMITRAZ, ACADREX, ACARAC, BAAM, BUMETRAN, DANICUT, ECTODEX, EDRIZAR, GARIAL, MITAC, METEX, OVIDREX, TAKTIC, TRIATOX, TRIAZID

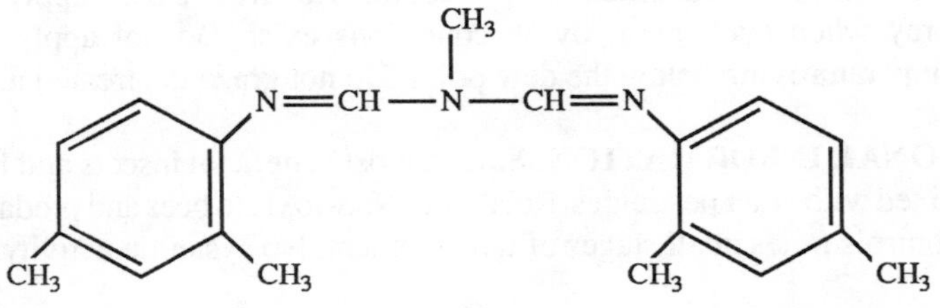

(N'-2,4-Dimethylphenyl)-N-[[(2,4-dimethylphenyl)imino] methyl]-N-methylmethanimidamide

TYPE: Amitraz is a triazapentadiene compound with insecticidal and acaricidal activity.

ORIGIN: The Boots Company, Ltd., 1971. Licensed to be sold in the U.S. by Nor-Am Chemical Co.

TOXICITY: LD_{50}-523 mg/kg.

FORMULATIONS: 25% and 50% WP, 12.5 and 20 EC.

PHYTOTOXICITY: Crop injury to young peppers and pears under high temperature conditions has been reported.

USES: Pears and cattle. Outside the U.S. it is used on apples, pears, peaches, citrus, melons, cucumbers, cotton, cattle, sheep, pigs, goats, and dogs.

IMPORTANT PESTS CONTROLLED: Red spider mites, physlla, leaf miners, scale insects, aphids, and on cotton whitefly, jassids, eggs, first instar larvae of Lepidoptera, such as bollworms, leafworms, loopers, and perforators.

Ectoparasites of animals include ticks, mange mites, and lice (on cattle), ticks, mange mites, lice and keds (sheep and goats), mange mites and lice (pigs), and ticks and mange mites (dogs).

RATES: Crops-Applications of 20-60 g a.i./100 1 water, or .75-1.5 lb a.i./Acre.

Animals — As a spray or dip for cattle, wash containing up to 0.025% a.i. As a dip for sheep, wash containing up to 0.05% ai. As a spray for pigs and pig buildings, wash containing up to 0.1% ai.

APPLICATION: Apply to crops when pests are first seen, and repeat as necessary. Apply to animals when conditions or infestation levels warrant treatment.

PRECAUTIONS: Harmful to fish. Do not store the EC below 32°F. Noncompatible with alkaline compounds, parathion, Cyprex, and others. Mixtures with parathion must also be avoided for application to top fruit. Do not apply as a summer spray when cool, poor, dying conditions exist. Do not apply when night temperatures are below the dew point. Do not graze the treated areas.

ADDITIONAL INFORMATION: Safe to most beneficial insects and bees. Can be mixed with other pesticides. Relatively Non-toxic to bees and predatory insects. Controls mites in all stages of development. No systemic activity.

NAMES

BENZOMATE, BENZOXIMATE, AAZOMAT, ARTABAN, CITRAZON

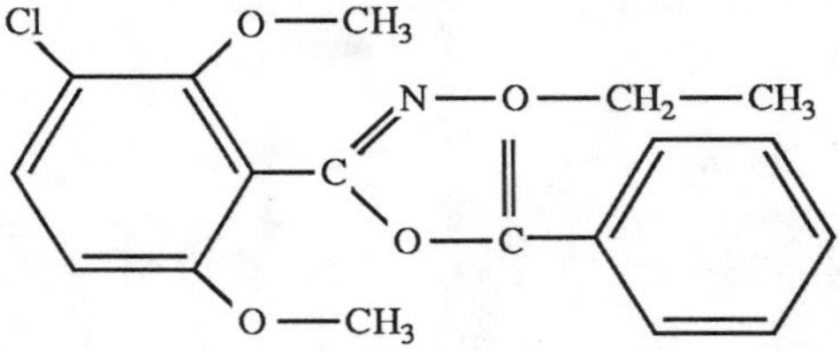

Ethyl O-benzoyl 3-chloro-2,6-dimethoxy-benzohydroximate

TYPE: Benzomate is a chlorinated compound used as a contact and residual acaricide.

ORIGIN: Nippon Soda Co. Ltd. of Japan, 1971.

TOXICITY: LD_{50}-15,000 mg/kg.

FORMULATION: 20% EC.

PHYTOTOXICITY: Non-phytotoxic when used as directed.

USES: In Japan and other countries, on citrus, apples, pears, plums, and peaches, and in Europe, on grapes and apples.

IMPORTANT PESTS CONTROLLED: Mites, both the eggs and adults.

APPLICATION: Apply to foliage when mites first appear, and repeat as necessary.

PRECAUTIONS: Not for sale or use in the U.S. Do not mix with EPN or Bordeaux mixture. Somewhat toxic to fish.

ADDITIONAL INFORMATION: Almost harmless to beneficial insects and predators. Gives both quick knockdown and residual control of mites. Compatible with most other pesticides. Considered to be effective by contact, as a stomach-poison, and on the eggs of mites.

NAMES

PLIFENATE, BAYGON-MEB, MEB-6046

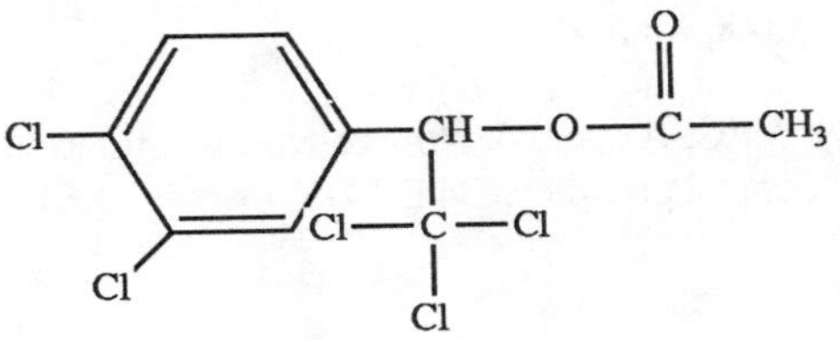

2,2,2-Trichloro-1-(3,4-dichlorophenyl)-ethanol acetate

TYPE: MEB-6046 is an organic compound used as a contact insecticide.

ORIGIN: Bayer AG of Germany, 1974.

TOXICITY: LD_{50}-10,000 mg/kg.

FORMULATIONS: Aerosols in combination with other pesticides. 30% WP.

PHYTOTOXICITY: Not for use on plants or crops.

IMPORTANT PESTS CONTROLLED: Mosquitoes, houseflies, lice, bed bugs, carpet beetles, clothes moths, and others.

USES: Household and industrial insect control.

APPLICATION: Apply as a contact spray, or to surfaces where insects will settle.

PRECAUTIONS: Not for sale or use in the U.S. Toxic to fish.

ADDITIONAL INFORMATION: Remains effective for several weeks. To be used primarily by pest-control operators. Sold mixed with permethrin or dichlorvos.

NAMES

METHOPRENE, ALTOSID, APEX, DIACON, DIANEX, KABAT, MANTA, MINEX, PHAROID, PRECOR, ZR-515

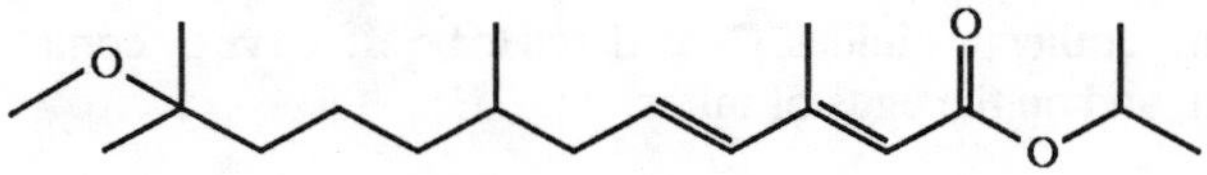

Isopropyl (2E,4E)-11-methoxy-3,7,11-trimethyl-2,4-dodecadienoate

TYPE: Altosid is an organic compound being used as an insect growth regulator.

ORIGIN: Zoecon Corporation, (Div. of Sandoz Crop Protection) 1973.

TOXICITY: LD_{50}-34,600 mg/kg.

FORMULATIONS: 5 EC. A special, slow release formulation has been developed for mosquito-control granules. .34 EC, 4% briquet, .86 lb/gal EC. Ant baits are being used.

PHYTOTOXICITY: None reported to date.

USES: Mosquito control in noncrop areas, rice, and pastures. Fed to livestock to control flies in the manure. Also used on stored tobacco and peanuts, and to eradicate Pharaoh's Ants. Used in mushroom houses, tobacco warehouses, and food processing plants. Used on chrysanthemum grown in greenhouses. Used as a feed through insecticide on livestock.

IMPORTANT PESTS CONTROLLED: Ants, cigarette beetle, cucumber beetle, fleas, flies, leafhoppers, leaf miners, lice, mosquitoes, moth, tobacco moth, and others.

RATES: 1/40-1/50 lb a.i./A.

APPLICATION: Mosquitoes — Apply at 2nd, 3rd, and 4th instar larvae stage. Flies — Free-feed to cattle and poultry. May be applied by air. Applied to tobacco as it goes into storage. For ant control, apply as a bait. Used for indoor flea control.

PRECAUTIONS: Somewhat toxic to fish. Do not mix with oil. Shrimp and crabs may be killed. Do not combine with other pesticides.

ADDITIONAL INFORMATION: Especially effective against flies. Exhibits morphological, rather than toxic activity against certain insect species. Stable, but nonpersistent. Relatively Non-toxic to nontarget species. Effective at very low rates. Growth regulators such as these prevent the insects from maturing to the adult stage of growth so they cannot reproduce. Shows little or no effect of the adult or pupae stage of insect development. Results are only seen by the number of emerged adults. Used in Japan to treat silkworms, making them extend the time period in which they produce silk.

NAMES

HYDROPRENE, ALTOZAR, GENCOR

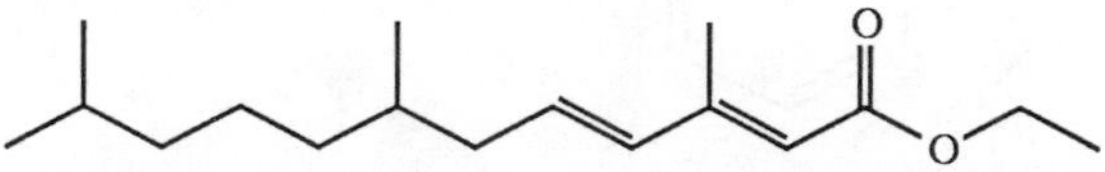

Ethyl (2E,4E)-3,7,11-trimethyl-2,4-dodecadienoate

TYPE: Hydroprene is a synthetic organic compound with juvenile hormone activity that is used as an insect growth regulator. The compound is classified by EPA as a biochemical biorational pesticide.

ORIGIN: Zoecon Corporation, (Div. of Sandoz Crop Protection) 1971.

TOXICITY: LD_{50}-5,100 mg/kg.

FORMULATIONS: 65.7% Emulsifiable Concentrate, 15.0% Hydroprene Concentrate for Formulating Use Only, 1.2% Total Release Fogger, 0.6% Total Release Fogger, 0.6% Aerosol.

PHYTOTOXICITY: Not applicable to current indoor uses.

USES: Within residential buildings including homes and apartment buildings and within nonfood areas of industrial, institutional, and commercial buildings.

IMPORTANT PESTS CONTROLLED: Cockroaches.

RATES & APPLICATIONS: EC and Aerosol applied as a spot or crack and crevice treatment. Foggers are applied at a rate of 2 oz/1000 $ft.^3$

ADDITIONAL INFORMATION: Applications of hydroprene to a population of roaches has no immediate effect since it exhibits very low direct toxicity. Adult insects remain unaffected and immature insects continue to develop through successive instars. Roaches in contact with hydroprene prior to the last ten days of the last nymphal instar emerge morphologically deformed (grotesquely shaped wings and a very dark exoskeleton) and physiologically unable to reproduce.

NAME

JH-388

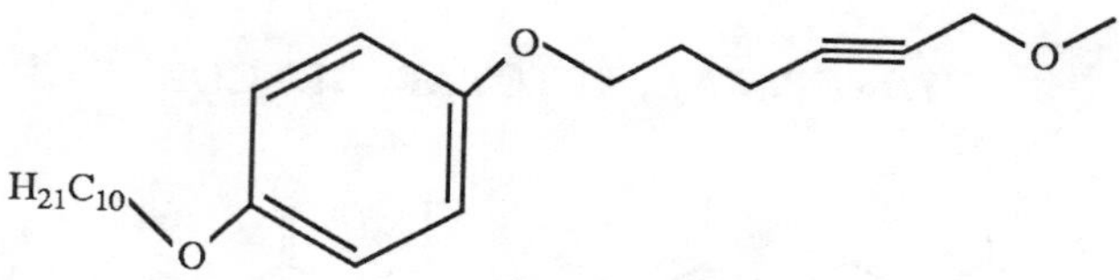

$H_{21}C_{10}$ O O O

1-Decyloxy-4-[(7-oxa-oct-4-ynyl)]-oxybenzene

TYPE: JH-388 is an organic compound used as an acaricide-ovicide.

ORIGIN: Farmoplant of Italy, 1981.

TOXICITY: LD_{50}-5000 mg/kg.

FORMULATION: 20% EC.

PHYTOTOXICITY: Non-phytotoxic when used as directed.

IMPORTANT PESTS CONTROLLED: Mite eggs.

USES: Experimentally being used on fruit crops and field crops.

RATES: Applied at 3 L (20 EC)/ha or at 150-250 g (20 EC)/100 L of water.

APPLICATION: Applied as a foliar spray to field crops and a dormant spray and summer spray to fruit crops.

PRECAUTIONS: Used on an experimental basis only. No activity on adult mites.

ADDITIONAL INFORMATION: May be tank mixed with additional acaricides or with oil.

NAME

JH-286

$O-CH_2-CH_2-CH_2-C \equiv C-Cl$

1,(5-Chloro-pent-4-ynyl)-oxy-4-phenoxy-benzene

TYPE: JH-286 is an organic compound used as a juvenile hormone insecticide, growth regulator.

ORIGIN: Farmoplant of Italy, 1981.

TOXICITY: LD_{50}-5800 mg/kg.

FORMULATION: 5 & 10% solution, 1% bait.

PHYTOTOXICITY: Non-phytotoxic when used as directed.

IMPORTANT PESTS CONTROLLED: Active against ants, fire ants, mosquito larvae, fleas, scales, sciarid flies, and others.

USES: Experimentally being tested as a juvenile growth hormone on the above insects.

APPLICATION: Apply where the immature forms of the insects will come in contact with the material. Used as a bait or as a spray.

PRECAUTIONS: Used on an experimental basis only. No effects on the adult insects.

ADDITIONAL INFORMATION: The hormone action causes developing abnormalities in the immature insects. Extremely selective in activity.

NAMES

TRIFLUMURON, ALSYSTIN, BAY SIR 8514

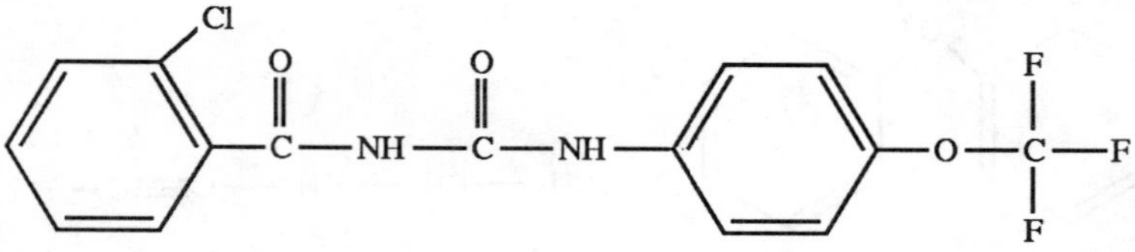

2-Chloro-N-[[[4-(trifluoromethoxy) phenyl]-amino]carbonyl]benzamide

TYPE: BAY SIR 8514 is a benzoylurea compound used as a nonsystemic insecticide-growth regulator with stomach poison activity.

ORIGIN: Bayer AG of Germany, 1978.

TOXICITY: LD_{50}-5,000 mg/kg.

FORMULATIONS: 25% WP, 250 SC. .25WP, .065 EC 4 lb flowable.

PHYTOTOXICITY: Non-phytotoxic when used as directed.

USES: Used on corn, cotton, soybeans, coffee, rice, vegetables, livestock, mosquito control, fruit trees, premises, forest, and rangeland Outside the U.S.

IMPORTANT PESTS CONTROLLED: Armyworms, boll weevils, cabbage looper, Colorado potato beetle, codling moth, cotton leaf perforator, flies, gypsy moth. Heliothis spp., mosquitoes, pear pyslla, and others.

RATES: Applied at .05-.5 lb a.i./A.

APPLICATION: Apply at the first sign of infestation and repeat as necessary. Enters the insect through the feeding tract. Toxic to fish.

PRECAUTIONS: The EC formulation appears more active than the WP. Not for sale or use in the U.S.

ADDITIONAL INFORMATION: This is an insect growth regulator that is slow-acting, moderately residual, and easy on predators. Use on chewing insects in their larval stages. Considered a chitin inhibitor and intervenes with the moulting process of the growing larvae. Also, egg sterility results from the female's exposure to treated surfaces. Nonsystemic. Poor activity against sucking insects and spider mites. Some activity as a contact insecticide. Should be applied earlier than conventional insecticides. After taking up the active ingredient the larvae fail to survive the next molt. Effectiveness increases with temperatures. May be mixed with other pesticides.

NAMES

***FLUBENZIMINE*, BAY-SLJ-0312, CROPOTEX**

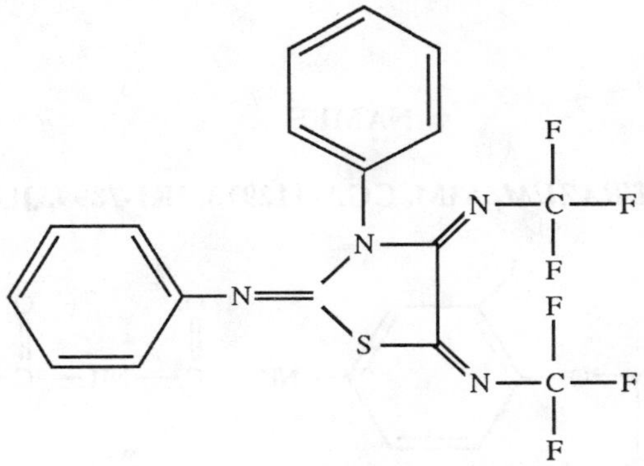

N-[3-Phenyl-4,5-bis[(trifluoromethyl)imino]-2-thiazolidinylidene]benzenamine

TYPE: Flubenzimine is a benzenamine compound used as an acaricide growth regulator.

ORIGIN: Bayer Ag of West Germany, 1981. Being developed in the U.S. by Mobay Chemical Corp.

TOXICITY: LD_{50}-2685 mg/kg. May cause eye and skin irritation.

PHYTOTOXICITY: Satsuma plums have been injured on the young leaves.

FORMULATION: 50% WP.

IMPORTANT PESTS CONTROLLED: Mites.

USES: Sold outside the U.S. on pome fruits and citrus. Experimentally being tested on apples, tea, vegetables, almonds, cotton, beans, and other crops.

RATES: Applied at .05% ai concentration.

APPLICATION: Apply when first mites appear. Early application gives the best results.

PRECAUTION: Used on an experimental basis only. No ovicidal effects. Slow acting. Toxic to fish.

ADDITIONAL INFORMATION: The effects are similar to those of chitin inhibiting insect growth regulators. Treated mobile stages develop to the next nonmobile stage, but then do not emerge. Treated adults produce fewer eggs. Excellent residual activity. May be mixed with other pesticides. Slow acting.

NAMES

***CHLORFLURAZUM*, AIM, CGA-112913, IKI-7899, JUPITER**

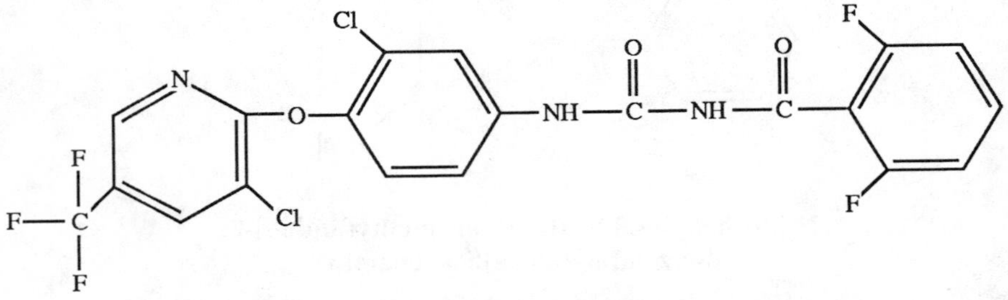

N-[4-(3-Chloro-5-trifluoromethyl-2-pyridinyl-oxy)-3,5-dichloro-phenyl-aminocarbonyl]-2,6-difluoro-benzamide

TYPE: CGA-112913 is a substituted urea compound used as an insect growth regulator.

ORIGIN: CIBA-Geigy and Ishihara of Japan, 1982.

TOXICITY: LD_{50}-1000 mg/kg.

FORMULATIONS: 25% WP, .05% EC, 500 g a.i./L flowable.

PHYTOTOXICITY: Non-phytotoxic when used as directed.

IMPORTANT INSECTS CONTROLLED: Bollworms, budworms, armyworms, leafworms, loopers, boll weevil, and others.

USES: Experimentally being tested on cotton.

RATES: Applied at .12-.25 lb a.i./A.

APPLICATION: Apply when Heliotis reach a 10-15% larval infestation. Make 4-6 applications as necessary. For boll weevil, application should begin at pinhead square stage of cotton growth. Make 4-6 applications at 5-14 day intervals.

PRECAUTIONS: To be used on an experimental basis only. Not effective against aphids, whiteflies, or mites.

ADDITIONAL INFORMATION: Disrupts the chitin formation in the insect. Has both contact and stomach-poison activity. No systemic activity. No effect on adult insects, however, the eggs fail to hatch on treated insects. Treated larvae fail to molt correctly, causing death.

NAMES

***CLOFENTEZINE*, ACARISTOP, APOLLO, APOLO, NC-21314, PANATAC**

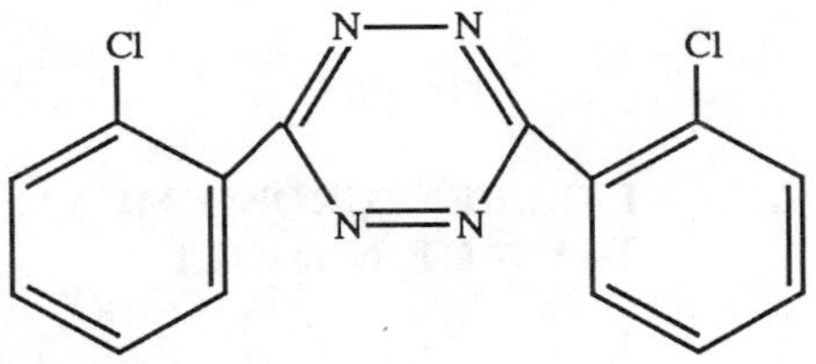

Cl N—N Cl N═N

3,6-bis(2-Chlorophenyl)-1,2,4,5-tetrazine

TYPE: Clofentezine is a diaryl tetrazine compound used as a contact acaricide.

ORIGIN: FBC Ltd. of England, 1981. Being developed in the U.S. by Nor-Am Chemicals and Schering AG outside the U.S.

TOXICITY: LD_{50}-1332. May cause eye and skin irritation.

FORMULATIONS: 4 lb/gal. suspension concentrate; 50% wettable powder.

PHYTOTOXICITY: Slight injury to roses grown under glass has been reported.

USES: Sold on apples, pears, peaches, citrus, cotton, cucumbers, melons, peppers, vines, and ornamentals outside the U.S. Experimentally being used on these crops in the U.S.

IMPORTANT PESTS CONTROLLED: Mites.

RATES: 20-50 g a.i./100 L water. Apply at 2-4 oz. a.i./Acre.

APPLICATION: To control European red mite, apply pre-blossom before winter egg-hatch or early post-blossom before mite numbers exceed 1 per leaf. If a repeat application is necessary, apply before mite numbers exceed 3 per leaf. For two-spotted mite and related species make the first application before mites exceed 1 per leaf and repeat as above.

PRECAUTIONS: The product may leave a slight pink deposit which may be noticeable on the petals of white or other pale colored flowers. Toxic to fish.

ADDITIONAL INFORMATION: Clofentezine is a specific acaricide which acts primarily as an ovicide/larvicide; it is particularly effective against winter eggs of European red mite. Following early season applications, it gives excellent residual control. It is safe to bees, beneficial insects, and predatory mites. Not effective against adult mites. Up to 80 days control has been noted. Slow acting. Compatible with other pesticides.

NAMES

***TEFLURON*, TEFLUBENZURON, CME 134, DART, DIARACT, NOMOLT**

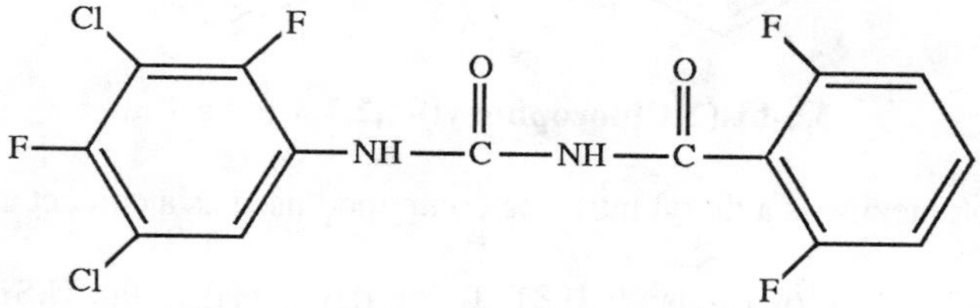

1-(3,5-Dichloro-2,3-difluorophenyl)-3-(2,6-difluorobenzoyl)-urea.

TYPE: CME 134 is a new insect growth regulator belonging to the group of benzoyl-urea compounds.

ORIGIN: Celamerck GmbH & Co. of West Germany, 1980.

TOXICITY: LD_{50}-5000 mg/kg.

FORMULATION: SC: 150 g/L.

PHYTOTOXICITY: Non-phytotoxic when used as directed.

USES: Used on pome fruit, citrus, cabbage, grapes, potatoes, forestry, cotton, soybeans, maize, sorghum, and others outside the U.S.

IMPORTANT INSECTS CONTROLLED: Alfalfa weevils, boll weevils, codling moth, Colorado potato beetle, diamond back moth, European corn borer, fall armyworm, grape berry moth, gypsy moth, hornworms, loopers, mosquitoes, pear psylla, pine sawfly, spruce budworms, and many others.

RATES: Applied at 0.01-0.1 lb/A.

APPLICATION: Apply when first larvae is visible. Repeat if necessary.

ADDITIONAL INFORMATION: Lacks fast initial effectiveness but possesses excellent residual activity. It interferes with the chitin synthesis. It may influence the fertility of female insect after contact or ingestion. It is nonsystemic, good coverage of the foliage is therefore recommended. Safe to many beneficial insects.

NAMES

FLUCYCLOXURON, ANDALIN, PH 70-23

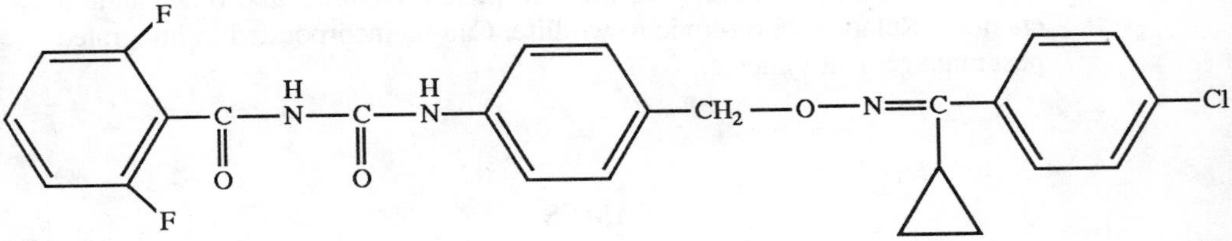

1-[a-(4-chloro-a-cyclopropylbenzylideneamino-oxy)-p-tolyl]-3-(2,6-difluorobenzoyl)urea

TYPE: Benzoylurea-type acaride/insecticide interfering with chitin deposition.

ORIGIN: DUPHAR B.V., The Netherlands; 1983 research discovery. In development stage. Registration applied for in various countries.

TOXICITY: LD_{50} >5000 mg/kg bw. Not mutagenic or teratogenic.

FORMULATIONS: Liquid 25%.

PHYTOTOXICITY: Non-phytotoxic at recommended rates and uses.

USES: Being tested world-wide on various pome and citrus fruit crops, nuts, field crops, vegetables and ornamentals.

IMPORTANT PESTS CONTROLLED: Most species of spider mites and rust mites. Important insects controlled include armyworms, bollworms, most cabbage caterpillars, codling moth, rice leaffolder, various leafrollers, Colorado potato beetle, most soybean caterpillars, mosquitoes and many others.

RATES: Applied at 0.0625-0.125 lb a.i./A.

APPLICATIONS: To control European red mite, apply at 75% winter egg hatch (around petal fall). For other spider mites and rust mites apply at the start of population build-up, before appearance of adults. If many adults are present, combination with an adulticide is recommended. To control insects, apply around oviposition time of adults or at early larval instar stages.

PRECAUTIONS: No effects on adult mites or insects. Toxic to crustaceans. No EPA-approved uses as yet.

ADDITIONAL INFORMATION: This product interferes with the formation of the insect's cuticle. Active on the larval stages of development, causing an inability to moult successfully. Feeding will continue for a short time after application (until the next moult), so results may not be visible immediately. Has a long residual activity. Relatively harmless to predatory mites and other natural enemies. Relatively non-toxic to wildlife. Can be incorporated in integrated pest management systems.

NAMES

***CYROMAZINE*, ARMOR, CGA-72662, CITATION, LARVADEX, NEPOREX, TRIGARD, VETRAZIN**

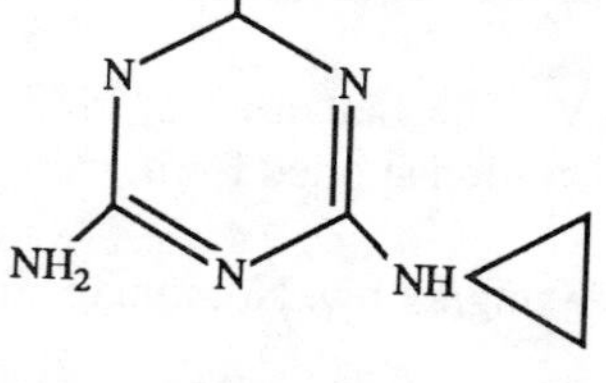

NH_2 N N NH_2 N NH

N-Cyclopropyl-1,3,5-triazine-2,4,6-triamine

TYPE: Cyromazine is a trizine compound used as an insect growth inhibitor.

ORIGIN: Ciba Geigy Corp., 1979.

TOXICITY: LD_{50}-3387 mg/kg. Slightly irritating to the eyes and skin.

FORMULATIONS: .3% premix, 5% soluable concentrate. 75 WP, 50% SP.

USES: Used on celery, lettuce,and to control flies in manure and other areas by feeding to poultry or treating the surface areas. Sold for fly control on animals in some countries.

IMPORTANT INSECTS CONTROLLED: Flies, leafminers.

RATES: Feed to poultry at .125-.5 mg/kg. As a surface spray, apply at .05% concentrate in .5-1 gal of water/100 sq ft. On crops apply at 1/6 lb, formulation.

APPLICATION:

1. Surface spray-Spray over the manure surface and repeat at 14-day intervals.
2. Premix-Begin feeding when adult flies become active, and continue through the season.
3. Foliar Spray-Apply when insects appear and repeat as necessary. Apply by ground or air.

PRECAUTIONS: Do not apply directly onto livestock.

ADDITIONAL INFORMATION: Contact activity against the larvae of most fly species. Fly control is observed 2-3 days after treatment is started, and will continue for 2-3 days after it has stopped. Appears to be selective toward Dipterous species. Controls fly larvae in manure after being fed to poultry.

NAMES

HYDRAMETHYLNON, AMIDINOHYDRAZONE,
PYRAMDRON, AC 217-300, AMDRO, COMBAT, CYAFORGEL, MAXFORCE, WIPEOUT

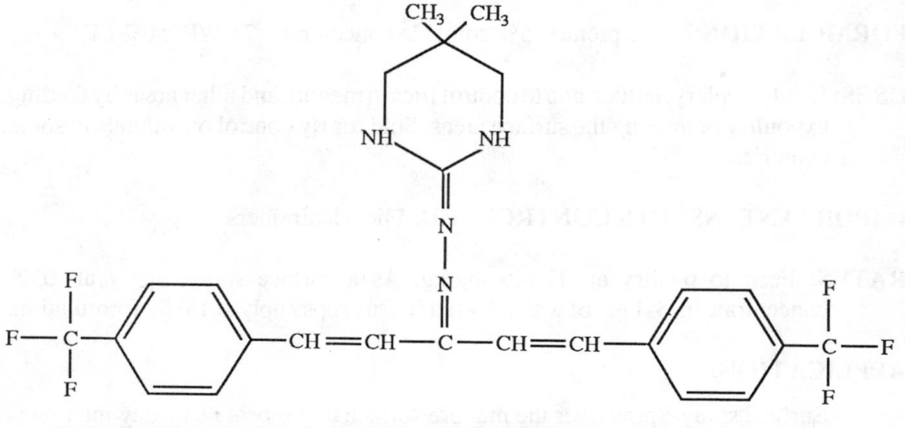

Tetrahydro-5,5-dimethyl-2(1H)-pyrimidinoine(3-(4-(trifluoromethyl) phenyl)-1-(2-(4-(trifluoro-methyl)phenyl)ethenyl)-propenylidene)hydrazone

TYPE: Amdro is an organic compound used as a slow-activating stomach-poison insecticide.

ORIGIN: American Cyanamid, 1979.

TOXICITY: LD_{50}-1131 mg/kg. Irritating to the eyes.

FORMULATIONS: .88% bait utilizing soybean oil as the attractant on inert corn grit carriers. Also a 1.65% bait for cockroaches.

PHYTOTOXICITY: Non-phytotoxic when used as directed.

USES: Use on pastures, rangegrass, lawns, turf, and noncrop areas.

IMPORTANT PESTS CONTROLLED: Fire ants and cockroaches.

RATES: The bait is applied at 1-1.5 lb/A.

APPLICATION: Apply where ants have their nests. Also apply as a broadcast treatment. Apply when ants are active and soil temperatures are above 60°F. Re-treat in 4 months, if necessary. For spot treatment, distribute 3-4 ft around the base of the mound. For roach control use 4-6 feeding stations/100 sq. ft.

PRECAUTIONS: Toxic to fish. May attract rodents and domestic animals.

ADDITIONAL INFORMATION: Does not bioaccumulate in the environment. Nonsystemic in plants. Not expected to be toxic to fish in the natural environment due to its low water solubility and rapid degradation in sunlight. Allow 2-4 weeks for control. May be applied by air. Controls roaches for 2-3 months.

NAMES

***BENSULTAP*, BANCOL, RUBAN, TI-78, VICTENON, ZZ-DORICIDA**

S,S-[2-(dimethylamino)trimethylene]bis(benzenethiosulfonate

TYPE: Bensultap is an organic compound used as a contact and stomach-poison insecticide.

ORIGIN: Takeda Chemical Ind. of Japan, 1978.

TOXICITY: LD_{50}-1105 mg/kg.

FORMULATIONS: 50% WP. 4% granules.

PHYTOTOXICITY: Some phytotoxicity was noted against certain varieties of apples, peaches, and citrus.

IMPORTANT PESTS CONTROLLED: Colorado potato beetle, thrips, leafrollers, stem borers, cabbageworm, armyworms, leaf miner, boll weevil, aphids, and others.

USES: Outside the U.S. on corn, wheat, sugarbeets, tea, potatoes, grapes, rice, vegetables, fruit trees, and other crops.

RATES: Applied at .30-.75 kg/ha.

APPLICATION: Apply when insects appear and repeat as necessary.

PRECAUTIONS: Not for sale or use in the U.S. Do not mix with alkaline compounds.

ADDITIONAL INFORMATION: The discovery of this product was based on research done on a naturally occurring substance, Nereistoxin, which was found in certain marine annelids. Coleopterous and Lepidopterous insects are primarily controlled. Relatively safe to fish. Activity is totally different than other insecticides. This chemical is not a cholinesterase inhibitor, but as a synaptic blocking agent for the insects central nervous system. Treated insects show sluggish movement and discontinue feeding followed later by death. Slow acting.

NAMES

***THIANITRILE*, PACT, SN 72129**

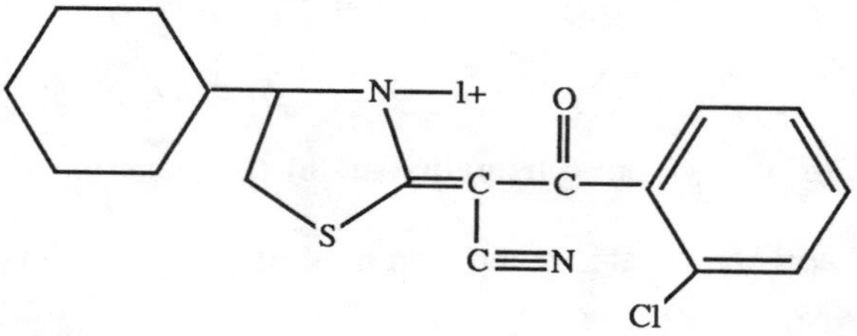

2-Chloro-beta-oxo-alpha-(4-phenyl-2-(3*H*)-thiazolylidene)-benzenepropanenitrile

TYPE: SN-72129 is a new class of compounds that is used as a contact and stomach-poison insecticide.

ORIGIN: Schering of West Germany, 1982. Being developed in the U.S. by Nor-Am Chemical Co.

TOXICITY: LD_{50}-5000 mg/kg. May cause eye irritation.

FORMULATION: 50% WP.

PHYTOTOXICITY: Non-phytotoxic when used as directed.

USES: Experimentally being tested on potatoes, alfalfa, pears, and stored grains.

IMPORTANT INSECTS CONTROLLED: Several species of Coleoptera, Homoptera, and Lepidoptera insecticide are controlled.

RATES: Applied at .05-.5 lb a.i./A.

APPLICATION: Apply when insects appear and repeat as necessary.

PRECAUTION: Used on an experimental basis only.

ADDITIONAL INFORMATION: No systemic activity. Ineffective against mites and aphids and very selective towards beneficial insects.

NAMES

NC-129

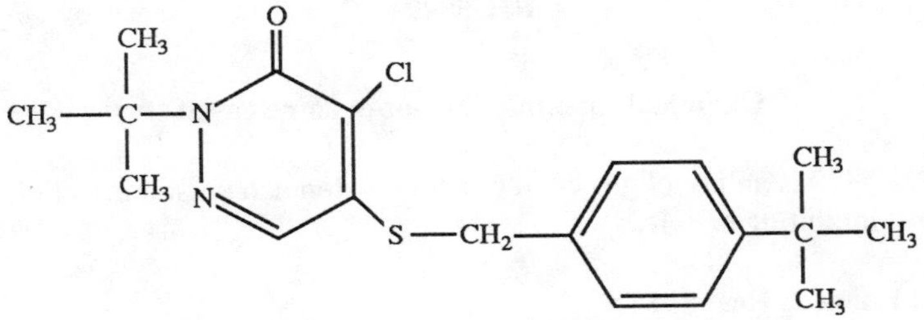

4-Chloro-2(1,1-dimethylethyl)-5-[[[4-(1,1-dimethyl-ethyl) phenyl]methyl]thio]-3(2H)-pyridazinone

TYPE: NC-129 is a pyridazinone compound used as a selective miticide-insecticide.

ORIGIN: Nissan Chemical Industries of Japan 1985.

TOXICITY: LD_{50} 435 mg/kg. May cause slight eye irritation.

FORMULATIOS: 15% EC, 20% WP.

PHYTOTOXICITY: Non-phytotoxic when used as directed. Injury has been reported on eggplant.

IMPORTANT PESTS CONTROLLED: Mites, thrips, aphids, leafhoppers, whiteflies and others.

USES: Experimentally being tested on citrus, fruit crops, grapes, vegetables, cotton, tea, ornamentals and others.

RATES: Applied at 50-300 ppm or at .2-.5 kg/ha.

APPLICATION: Apply when insects appear and repeat as necessary.

PRECAUTIONS: To be used on an experimental basis only. Thorough coverage is necessary for good control. Not effective on Lepidoptera, Coleoptera, Hymenoptera or Dipterous insect species.

ADDITIONAL INFORMATION: Effective on all stages of mites. Temperatures do not seem to effect the activity. Non-systemic. Quick acting to motile mite stages. Good residual activity.

NAME

RH-5849

Chemical formula has not been released

TYPE: RH-5849 is an insect growth regulator — stomach poison insecticide with molt inhibiting activity.

ORIGIN: Rohm & Haas 1987.

TOXICITY: LD_{50} 435 mg/kg.

FORMULATIONS: 2 lb/gal. flowable.

PHYTOTOXICITY: Non-phytotoxic when used as directed.

INPORTANT INSECTS CONTROLLED: Codling moth, leafminers, oriental fruit moth, leafrollers, Colorado potato beetle, rice stemborer, cutworms, armyworms and others.

USES: Experimentally being tested on apples, potatoes, rice, vegetables and others.

RATES: Applied at .2-1 kg/ha.

APPLICATION: Apply just prior to egg laying and repeat at 14-21 day intervals.

PRECAUTIONS: Used on an experimental basis only. Not effective against aphids, mites, thrips or true bugs.

ADDITIONAL INFORMATION: Effects the larval states of Lepidoptera and some Coleoptera and Diptera insect species. Insects stop feeding within a few hours of treatment. Some ovicidal activity. Non-toxic to bees. Active on insects when applied to the soil. May be tank mixed with other pesticides.

NAME

RH-7988

Chemical formula not yet released

TYPE: RH-7988 is a highly selective fast acting systemic aphicide.

ORIGIN: Rohm & Hass 1986.

TOXICITY: LD_{50} 61 mg/kg. May cause eye and skin irritation.

FORULATIONS: 4EC, 2EC, 1% granules.

PHYTOTOXICITY: Some injury has been observed on alfalfa at high rates.

IMPORTANT PESTS CONTROLLED: aphids.

USES: Experimentally being tested on apples, citrus, fruit crops, cereals, sorghum, cole crops, lettuce, potatoes, sugar beets, tobacco, cotton, ornamentals and others.

RATES: Applied at 35-290 g/ha.

APPLICATION: Apply when insects appear and repeat as necessary. May also be applied to the soil.

PRECAUTIONS: Used on an experimental basis only. Toxic to fish.

ADDITIONAL INFORMATION: Effective on aphids only. Highly mobile within the plant. Fast acting. Good residual activity. May be applied with an oil based spray adjuvant for better control. Moves both upward and downward in the plant.

ORGANIC PHOSPHATES
Phosphoric Acid Prototypes

NAMES

PHOSPHAMIDON, APAMIDON, DIMECRON, DIMENOX, DIXON, FAMFOS, SWAT

```
                 O               Cl
CH3—O            ‖               |                  CH2 —CH3
      \          ‖               |                 /
       P—O—C═C—C—N
      /          |               ‖                 \
CH3—O           CH3              O                  CH2 —CH3
```

2-chloro-3-(diethylamino)-1-methyl-3oxo-1-propenyl dimethyl phosphate

TYPE: Phosphamidon is an organic-phosphate, systemic stomach-poison insecticide-acaricide.

ORIGIN: CIBA-Geigy, 1957.

TOXICITY: LD_{50}-17 mg/kg. Absorbed through the skin.

FORMULATIONS: 8 EC, 20% EC, 50% EC.

PHYTOTOXICITY: Injury has been reported on cherries, plums, peaches, and maple trees. Some injury to sorghum varieties related to Red Swazi. Apples and walnuts sometimes show a foliage tip burn, but this is not harmful to the crop.

USES: Apples, broccoli, cantaloupes, cauliflower, citrus, cotton, cucumbers, peppers, potatoes, tomatoes, walnut, and watermelon.

IMPORTANT PEST CONTROLLED: Aphids, stemborers, lygus bugs, leafhoppers, leaf miners, spruce budworm, beetles, thrips, codling moth, grasshoppers, mites, scale, bollworms, Mexican bean beetles, whiteflies, and many others.

RATES: Applied at 1/4 -1 1/2 lb a.i./A.

APPLICATION: Apply thoroughly when insects first appear. Repeat as necessary.

PRECAUTIONS: Workers should wear protective clothing when entering fields within 24 hours following application. Do not feed treated forage or crop residue to livestock. Incompatible with alkaline materials. Do not use or allow to drift on cherries. Highly toxic to bees. Do not mix with copper oxychloride. Do not combine with captan, folpet, or sulfur.

ADDITIONAL INFORMATION: A systemic compound, translocated through both the roots and the leaves. Noncorrosive. Does not accumulate in the soil. A selective compound with only slight contact and practically no ovicidal activity. Compatible with other pesticides Used on almost every crop grown in the world in the countries for which it is registered. Low toxicity to fish. Relatively safe to wild life. Shows little or no effect on many beneficial insects.

NAMES

DICROTOPHOS, BIDRIN, CARBICRON, DIAPADRIN, EKTAFOS

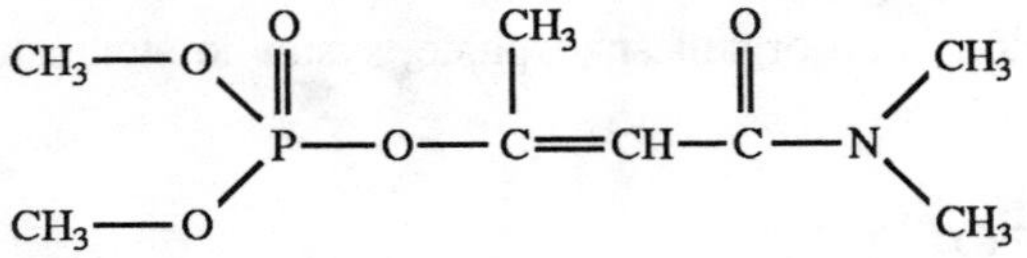

```
              O         CH3         O
CH3—O         ‖          |          ‖           CH3
      \       ‖          |          ‖          /
        P — O — C ═ CH — C — N
      /                                 \
CH3—O                                     CH3
```

Dimethyl phosphate ester of 3-hydroxy-N,N-dimethyl-cis-crotonamide

TYPE: Bidrin is an organic-phosphate insecticide-acaricide used as a contact and systemic stomach-poison.

ORIGIN: Shell Chemical Company and CIBA-Geigy, 1963.

TOXICITY: LD_{50}-17 mg/kg.

FORMULATIONS: 8 lb/gal soluble concentrate, 40 and 50% EC.

PHYTOTOXICITY: Considered nonphytotoxic when used at the recommended rates. It may be harmful to some varieties of grain seed.

USES: Cotton and pecans. Used outside the U.S. on these, as well as rice, potatoes, cereals, coffee, citrus, sugarcane, palms, tobacco, and for tick control in cattle.

IMPORTANT PESTS CONTROLLED: Aphids, mites, thrips, fleahoppers, grasshoppers, boll weevils, lygus bugs, rice stemborers, bollworms, leaf miners, stinkbugs, leafhoppers, and many others.

RATES: Applied at .1-.5 lb actual/A.

APPLICATION: Apply evenly at a uniform rate. Repeat as necessary. May be applied by ground or air. May be applied to ornamentals by the tree-injection method.

PRECAUTIONS: Birds and wildlife in treated areas may be killed. Do not apply during peak bee activity. Do not graze livestock on treated fields. Toxic to fish. May be corrosive to black iron, stainless steel type 304, and brass.

ADDITIONAL INFORMATION: Fast acting. Compatible with most other pesticides. More than 50% of the material is absorbed into the plant within 8 hours of application. 7-21 days' control can be expected.

NAMES

***MONOCROTOPHOS*, APADRIN, AZODRIN, BILOBRAN, CARBICRON, CRISODRIN, MONOCRON, NUVACRON, PANDAR, PHILLARDRIN, PLANTDRIN, SUSVIN**

$$(CH_3{-}O)_2P(=O){-}O{-}C(CH_3){=}CH{-}C(=O){-}CH{-}CH_3$$

Dimethyl phosphate of 3-hydroxy-N-methyl-cis-crotonamide

TYPE: Azodrin is an organic phosphate insecticide with systemic and contact activity.

ORIGIN: CIBA-Geigy Ltd., 1965.

TOXICITY: LD_{50}-8 mg/kg.

FORMULATIONS: 5 lb a.i./gal water miscible solution. 3.2 and 4.8 lb/gal soluble concentrates.

PHYTOTOXICITY: Considered nonphytotoxic at the recommended rates. Slight injury has been reported on certain apple, cherries, almond, peaches, and sorghum varieties.

IMPORTANT PESTS CONTROLLED: Bollworms, budworms, pink bollworms, loopers, lygus bugs, mites, boll weevils, and a number of others. Also a number of fruit and vegetable pests are controlled on crops outside the U.S.

RATES: Used at 1/2-lb actual/A.

USES: Cotton, citrus, sugarcane, peanuts, and ornamentals. Used outside the U.S. on fruit, grapes, vegetables, hops, citrus, olives, potatoes, tomatoes, tobacco, sorghum, sugar beets, rice, and ornamentals.

APPLICATION: Apply to give uniform coverage. Repeat as necessary. This may be every 5 days where infestations are heavy. Used as a seed treatment on cotton.

PRECAUTIONS: Highly toxic to birds and other wildlife. Toxic to bees. Somewhat corrosive to iron, steel, and brass. Do not graze livestock on treated fields. Do not handle nursery stock for 2 days after application. Do not store at temperatures below 40°F or above 80°F for prolonged periods.

ADDITIONAL INFORMATION: Contact and systemic activity. Penetrates the plant tissue rapidly. Compatible with most other pesticides, except those alkaline in reactions. Control should last for 15-20 days on many insects. Fast acting, with strong systemic activity.

NAMES

***CROTOXYPHOS*, CIODRIN, DECROTOX, DURAVOS**

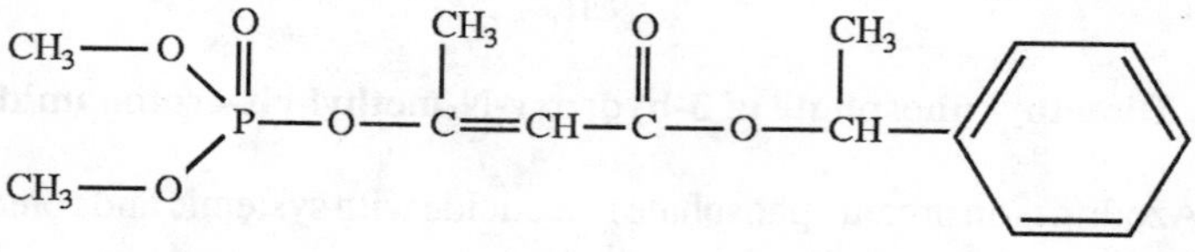

Dimethyl phosphate of alpha-methylbenzyl-3-hydroxy-cis-crotonate

TYPE: Ciodrin is an organic-phosphate insecticide developed for use on livestock.

ORIGIN: Shell Chemical Company, 1963.

TOXICITY: LD_{50} -21 mg/kg.

FORMULATIONS: 2, 3, and 4 EC, 2% solution, WP, and dusts.

PHYTOTOXICITY: Not generally applied to crops.

USES: Registered for use on cattle, sheep, goats, hogs, and agricultural premises, barns, fences, buildings, etc.

IMPORTANT PESTS CONTROLLED: Flies, lice, ticks, and mites.

RATES: Applied at 1-3 lb actual/100 gal of water.

APPLICATION:

1. Spray-Spray all parts of the animal with 2-4 qt of the spray solution. It may also be used to treat the animal premises. Repeat as necessary. May be applied daily as a fine mist. Also used by the pour-on method.
2. Backrubbers-Mix 1 gal of the EC in 50 gal of backrubber oil (diesel fuel or diesel plus motor oil). Check frequently.

PRECAUTIONS: Slightly corrosive to steel. Do not apply in excess of 2 oz/animal/day. Do not contaminate food or drinking water. Do not use on poultry houses. Toxic to fish.

ADDITIONAL INFORMATION: Compatible with other pesticides. Sometimes mixed with Vapona as an animal insecticide. Incompatible with most mineral carriers.

RELATED MIXTURES: CIOVAP — A cattle insecticide containing Ciodrin and Vapona, used to control flies. Developed by Shell Chemical Company.

NAMES

PROFENOFOS, CGA-15324, CURACRON, POLYCRON, SELECRON

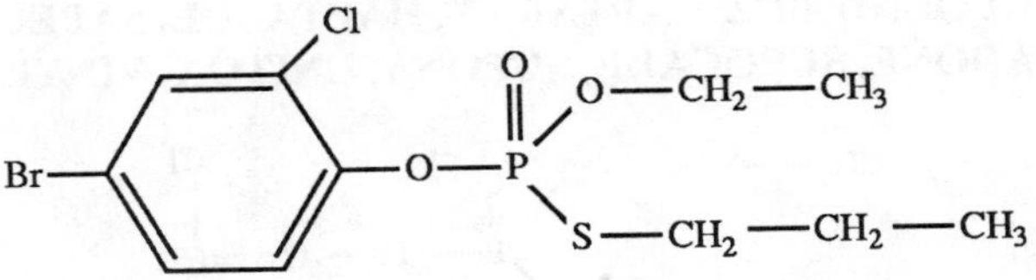

Cl
O
O—CH_2—CH_3
Br— —O—P
S—CH_2—CH_2—CH_3

_O_-4-Bromo-2-chlorophenyl _O_-ethyl S propyl-phosphorothioate

TYPE: Curacron is an organic-phosphate compound used as a contact and stomach-poison insecticide.

ORIGIN: CIBA-Geigy, Ltd., 1973.

TOXICITY: LD_{50}-358 mg/kg. Somewhat irritating to the skin and eyes.

FORMULATION: 6 EC, 40% EC, 60% EC.

PHYTOTOXICITY: Slight reddening to cotton has occurred.

IMPORTANT INSECTS CONTROLLED: Bollworms, boll weevils, fruit flies, locusts, mites, budworms, armyworms, leaf perforators, cabbage loopers, and others.

USES: Cotton. Used on cotton, potatoes, sugar beets, tobacco, soybeans, and vegetables outside the U.S.

RATES: Applied at .5-1 lb a.i./A.

APPLICATION: Use as a larvicide. Apply when insects appear, and repeat as necessary.

PRECAUTIONS: Toxic to fish, wildlife, and bees. Do not graze the treated land. Corrosive.

ADDITIONAL INFORMATION: A broad-spectrum insecticide. Nonsystemic. More effective on insects than mites. Most effective on lepidoptera insects. A larvicide that is most effective on the first to third instar larvae. May be applied by air. Has ovicidal properties.

NAMES

***CHLORFENVINPHOS,* APACHLOR, BIRLANE, COMPOUND 4072, HAPTARAX, HAPTASOL, SAPECRON, STEFADONE, SUPOCADE, SUPONA, UNITOX, VINYLPHATE**

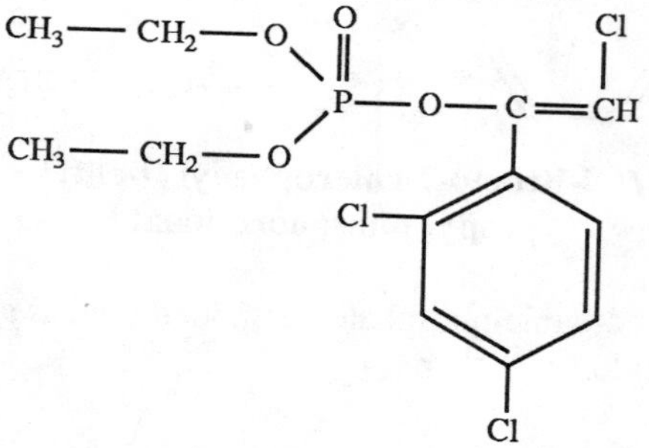

2-Chloro-1-(2,4-dichlorophenyl) vinyl diethylphosphate

TYPE: Chlorfenvinphos is an organic-phosphate, contact insecticide, with long residual effectiveness.

ORIGIN: Shell Chemical Co. and CIBA-Geigy, 1963.

TOXICITY: LD_{50}-10 mg/kg.

FORMULATIONS: 10% granules, 2 EC, 25% WP, and 5% dust.

PHYTOTOXICITY: Non-phytotoxic when used as recommended. Some injury has occurred when applied directly on the seed of certain crops.

USES: Used outside the U.S. on potatoes, rice, corn, vegetables, and other crops. Widely used as a livestock insecticide.

IMPORTANT PESTS CONTROLLED: Root maggots, rootworms, cutworms, scales, leafhoppers, Colorado potato beetle, ticks, lice, screwworm, and fleas.

RATES: Used at 1/2 to 2 lb actual/A. On livestock used at .05-.1% concentration.

APPLICATION: Soil treatment-Apply during the planting operation behind the covering discs and in front of the press wheel. A banded application at the time of the last cultivation may also be applied. Also applied as a foliar spray and as a seed treatment. Used on livestock as a dip or spray.

PRECAUTIONS: Do not graze livestock on treated crops. Do not use immature corn for silage or grazing. May be corrosive to iron, steel, and brass. Toxic to fish. Not used in the U.S.

ADDITIONAL INFORMATION: Most effective against soil insects. Compatible with most other pesticides. Up to 2 months' residual activity may be expected. Applied to poultry droppings for fly control. No residual toxicity to bees. A contact insecticide. Compatible with most other pesticides. Gives 2-3 weeks control.

NAMES

***TETRACHLORVINPHOS*, APPEX, GARDCIDE, GARDONA, RABON, RABOND, STRIOFOS**

(S)-2-Chloro-1-(2,4,5-trichlorophenyl) vinyl dimethylphosphate

TYPE: Tetrachlorvinphos is an organic phosphate compound with contact and stomach-poison activity.

ORIGIN: Shell Chemical Co., 1966. Being marketed in the U.S. as a livestock insecticide by Fermenta.

TOXICITY: LD_{50}-4000 mg/kg.

FORMULATIONS: 2 EC, 50 WP, 75 WP, 4 lb/gal oil solution, 3% dust.

PHYTOTOXICITY: Russeting may occur on Golden Delicious apples under some conditions.

USES: Registered in the U.S. on cattle, hogs, horses, poultry, and agricultural premises. Used in livestock feeds as a feed through insecticide. Outside the U.S. it is used as a livestock insecticide as well as apples, pears, citrus, stone fruits, rape, cotton, ornamentals, corn, soybeans, sugarcane, sunflower, tobacco, grapes, vegetables, forestry crops, and others, as well as a stored product insecticide.

IMPORTANT INSECTS CONTROLLED: Flies, lice, flour beetles, ticks, animal mites, stores product beetles, leaf miners, codling moth, leaf rollers, pear psylla, plum curculio, fruit flies, peach twig borer, bollworms, stem borers, armyworms, budworms, tent caterpillar, and many others.

RATES: Use at .5-1.5 lb a.i./A.

APPLICATION:

1. Foliage-Apply when insects appear and repeat as necessary.
2. Livestock-To livestock apply by direct application, back rubbers, dust bags or as a feed through insecticide. To livestock premises, apply by direct application.

PRECAUTIONS: Potentiates with parathion, methyl parathion, and malathion. Do not mix with dodine or alkaline compounds. Toxic to bees and fish. Not available in the U.S. for crop usage.

ADDITIONAL INFORMATION: Relatively non-hazardous to wildlife. Nonsystemic in activity. Effective against the adult and larvae forms of insects. Gives up to three week control. Compatible with other pesticides.

NAMES

***DDVP, DICHLORVOS,* ATGARD, BENFOS, CANOGARD, CEKUSAN, DEDEVAP, DICHLORPHOS, DIVIPAN, EQUIGARD, HERKAL, KRECALVIN, LINDAN, MAFU, MARVEX, NERKOL, NOGOS, NUVAN, OKO, PHOSVIT, PIRAN, RITON, TASK, UNIFOSZ, VAPONA, VAPONITE**

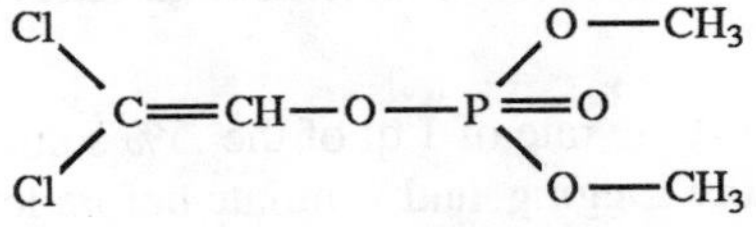

$$Cl \qquad\qquad\qquad\qquad O{-}CH_3$$
$$C{=}CH{-}O{-}P{=}O$$
$$Cl \qquad\qquad\qquad\qquad O{-}CH_3$$

2,2-dichlorovinyl dimethyl phosphate
hexahydro-4,7-methanoindene

TYPE: DDVP is an organic phosphate insecticide-acaricide effective as a fumigant, stomach, and contact poison.

ORIGIN: CIBA-Geigy and Bayer AG, 1960. Shell Chemical Company holds the U.S. patent. Produced by a number of manufactures today.

TOXICITY: LD_{50}-56 mg/kg. Absorbed through the skin.

FORMULATIONS: 1.4, 2, 4, and 10 EC, 10% aerosol, 50 and 90% concentrate solution, 93 % bait solution, .1-.5% liquid bait, .5% dry bait, 20% impregnated resin strip, 20% resin granules, 250, 500, & 1000 g/1 EC.

PHYTOTOXICITY: Non-phytotoxic when used as directed.

USES: Beef and dairy cattle, figs, goats, sheep, swine, poultry, mushroom houses, households, food-processing plants, outdoor fogging, and agricultural premises. Used outside the U.S. on apples, pears, peaches, rice, cotton, vegetables, grapes, tea, citrus, and other crops.

IMPORTANT PESTS CONTROLLED: Ants, aphids, mites, mealybugs, ticks, drosophila, centipedes, moths, cockroaches, roaches, crickets, fleas, flies, gnats, mosquitoes, sowbugs, spiders, wasps, and many others.

APPLICATION:

A. Livestock

1. Baits-Mix with a sweet bait (75% corn syrup and 25% water) to give a concentrate of .05%. Apply bait with a small brush to the faces of animals. Treat every morning for the first 2 weeks when the animals are in the barn. Apply as needed thereafter.

2. Sprays-Apply 1-2 oz of the 1% solution/animal as a mist spray daily. Apply so as not to wet the hair.

B. Agricultural Premises

1. Spray-Apply at the rate of one qt of the .5% solution/1000 sq ft. Remove animals before spraying. Ventilate before letting animals re-enter premises. Repeat as necessary. Also used in backrubbers or a back pour-on solution.

2 FOG-Apply at the rate of 1 qt of the .5% solution/18,000 cu ft. Remove animals before fogging, and ventilate before letting them re-enter.

3. Dry baits-Scatter at the rate of 1/4 lb/1000 sq ft for the control of houseflies.

4. Aerosols-Apply when the temperature is between 60-80°F. Place the cylinder in the enclosure 2-3 hours before using, so the temperatures will equalize. Point the nozzle upwards, walking one way toward the door through the building. Keep closed for 2 hours, and then ventilate for 2 hours.

PRECAUTIONS: Not accepted for use in dairy barns. Avoid direct application to livestock feed and water. Toxic to bees. Do not combine with alkaline compounds or with Morestan or Euparen. Toxic to fish.

ADDITIONAL INFORMATION: Used to a great extent in nonagricultural applications by the U.S. Department of Public Health for controlling annoying and harmful insects. Used to some extent as a fumigant in tobacco warehouses. Extremely fast knockdown effects. Used in many countries on crops just prior to harvest. Compatible with other compounds. Residual control of 2-3 weeks may be obtained.

NAMES

NALED, BROMCHLOPHOS, BROMEX, DIBROM, HIBROM

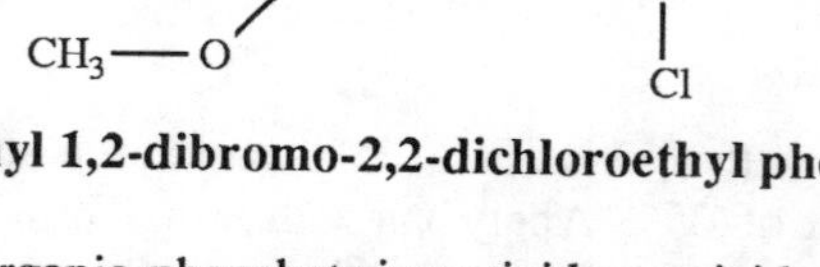

$$(CH_3O)_2P(=O)-O-CH(Br)-C(Br)(Cl)-Cl$$

Dimethyl 1,2-dibromo-2,2-dichloroethyl phosphate

TYPE: Dibrom is an organic-phosphate insecticide-acaricide which has both contact and stomach-poison activity with short residual effects.

ORIGIN: Valent Chemical Co.

TOXICITY: LD_{50}-430 mg/kg. Absorbed through the skin.

FORMULATIONS: 4, 8, 12, and 14 EC, 4% dusts.

PHYTOTOXICITY: Some injury has been reported on apples, pears, melons, cherries, plums, peaches, beans, cotton, and also on ornamentals such as White Butterfly Rose, Golden Rapture Rose, Green Wandering Jew, Dutchman's Pipe, Ornamental Cherries, Liquid Amber, Poinsettias, and Chrysanthemums. Do not apply to the Hegari variety of grain sorghum. It may also cause fruit spotting on nectarines. Do not use on Italia variety of grapes, or scarlet runner beans.

USES: Alfalfa, almonds, beans, broccoli, Brussels sprouts, cabbage, rangelands, cauliflower, celery, citrus, chard, collards, cotton, cucumbers, eggplant, grapes, hops, kale, lettuce, livestock, melons, oranges, peaches, peas, peppers, pumpkins, rice, pastures, soybeans, spinach, squash, strawberries, sugar beets, tobacco, tomatoes, turf, turnips, walnuts, watermelons, poultry, dogs, agricultural premises, and greenhouses.

IMPORTANT PESTS CONTROLLED: Loopers, corn earworms, mites, aphids, mosquitoes, houseflies, lygus bugs, armyworms, leafhoppers, leaf miners, cabbageworms, fleahoppers, bollworms, stinkbugs, cutworms, fruit flies, peach twig borers, spittlebugs, thrips, whiteflies, gnats, grasshoppers, and many others.

RATES: Applied at 1/8-3/4 lb actual/A in 40-100 gal of water.

APPLICATION: Begin as insects first appear. Repeat as necessary. Cover the foliage thoroughly. May be applied inside greenhouses to certain crops and ornamentals. For fly control, apply as a surface spray, space spray, or as a bait.

PRECAUTIONS: Do not apply when the temperature is over 90°F. Noncompatible with highly alkaline material such as lime and Bordeaux. Avoid application in periods of heavy bee activity. Will corrode iron, so clean equipment immediately after using. Do not store diluted spray. Irritating to the skin.

ADDITIONAL INFORMATION: Agitate while applying. Applied both by air and ground rig. Not toxic to fish at the levels used for the control of mosquito larvae. Fast knockdown of insects with a broad spectrum. Widely used as a public health insecticide for mosquito control.

NAMES

PROPAPHOS, KAYAPHOS, NK-1158

$$CH_3—CH_2—CH_2—O \quad (CH_3—CH_2—CH_2—O)P(=O)—O—C_6H_4—S—CH_3$$

4-(Methylthio) phenyl dipropyl phosphate

TYPE: Kayaphos is an organic phosphate compound used as a contact, systemic, and stomach-poison insecticide.

ORIGIN: Nippon Kayaku Co. of Japan, 1970.

TOXICITY: LD_{50}-70 mg/kg.

FORMULATIONS: EC, 20% dust, 5% granules.

USES: Paddy rice in Japan.

IMPORTANT PEST CONTROLLED: Most rice insect pests.

RATES: Applied at 600-800 g a.i./ha.

APPLICATIONS: Apply when insects appear and repeat as necessary. Also can be applied in the nursery box of rice planting machine.

PRECAUTIONS: Not for sale or use in the U.S. Toxic to fish.

NAMES

TRICHLORFON, ANTHON, BOVINOX, CEKUFON, CHLOROFOS, DANEX, DENKAPHON, DEP, DIPTEREX, DITRIFON, DYLOX, MASOTEN, NEGUVON, PROXOL, TRICHLORPHON, TRINEX, TUGON

$$(CH_3—O)_2P(=O)—CH(OH)—CCl_3$$

Dimethyl (2,2,2,-trichloro-1-hydroxy ethyl) phosphonate

TYPE: Dylox is a selective organic phosphate insecticide used on both plant foliage and livestock as a stomach-poison and as a contact insecticide.

ORIGIN: Bayer AG in Germany, 1952. Licensed to be sold in the U.S. by Mobay Chemical Corporation.

TOXICITY: LD_{50}-450 mg/kg.

FORMULATIONS: 40, 50, 80, & 95% soluble powder, 500 g/1 soluble concentrate, 250, 500, 700 g/1 ULV, 2.5 & 5% granules, 5% dust.

PHYTOTOXICITY: Sorghum and milo are severly injured. Injury has been reported on the foliage and the fruits of apples, and on carnations and zinnias. Cotton has shown injury if application is made while it is wet.

USES: Alfalfa, barley, blueberries, beans, beets, Brussels sprouts, cabbage, carrots, citrus, pasture and ranges, cauliflower, clover, collards, corn, cotton, cowpeas, flax, lima beans, oats, peppers, pumpkins, safflower, soybeans, sugar beets, tobacco, tomatoes, turf, wheat, ornamentals, agricultural premises, aquatic areas, poultry processing plants, and cattle. Also used as a public health insecticide and as a forest insecticide. Used outside the U.S. on these and other crops such as bananas, coffee, olives, mulberry, rice, grapes, tea, sugarcane, etc.

IMPORTANT PESTS CONTROLLED: Lygus bugs, flies, stinkbugs, cutworms, caterpillars, webworms, cockroaches, armyworms, leafhoppers, leaf miners, hornworms, and many others, as well as livestock-infesting insects such as ascarids, bots, pinworms, grubs, flies, screwworms, lice, ticks, fleas, and mites.

RATES: Applied at 1/4-2 lb actual/A. To livestock, apply 1/2 oz/100 lb body weight as a pour on.

APPLICATION:

1. Foliage-Apply evenly at a uniform rate. Repeat as necessary.
2. Premises and public health-Used in 1% active sugar fly baits, either liquid or dry. Do not contaminate milking equipment, feed, or water. May also be applied as a spray.
3. Livestock-Neguvon is the trade name of the livestock formulation. Applied to livestock by the pour-on technique, or by spraying.

PRECAUTIONS: Do not treat animals less than 3 months old nor treat lactating dairy cattle. Do not use on walls, floors, etc. previously treated with lime, whitewash, or other alkaline materials. Do not treat portions of building that animals may lick. Incompatible with alkaline materials and oils. Do not apply to beef cattle within 14 days of slaughter. Do not use on greenhouse crops. May cause spotting to automobile paint surfaces. Toxic to bees. Not too effective for the control of aphids, cabbage loopers, or mites.

ADDITIONAL INFORMATION: Applied directly to plant foliage without the need of emulsifiers or wetting agents. DDT-resistant flies and chlordance-resistant cockroaches are controlled, as well as some soil insects. Sometimes applied as a barrier strip around fields. Harmless to bees once it has dried on the plant. Agitation is not required once it is in solution. Compatible with most other pesticides. Does not significantly effect most beneficial insects. Used on poultry manure piles to control fly larvae.

NAMES

***ISOFENPHOS*, AMAZE, BAY-SRA-12869, OFTANOL, PRYFON**

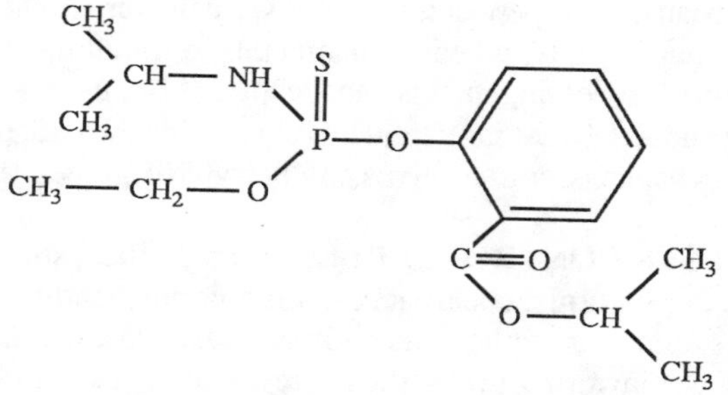

1-Methylethyl 2-((ethoxy((1-methylethyl)amino) phosphinothioyl)-oxy) benzoate

TYPE: Oftanol is an organic-phosphate compound used as a selective, soil insecticide.

ORIGIN: Bayer AG of Germany, 1975. Being marketed in the U.S. by Mobay Chemical Co.

TOXICITY: LD_{50}-32 mg/kg. Rapidly absorbed through the skin.

FORMULATIONS: 6 EC, 1.5%, 5%, and 20% granules, 2 EC seed-dressing powder, mixtures with Thiram, 40% WP.

PHYTOTOXICITY: Non-phytotoxic when used as directed.

USES: Turf and used to control termites in structures. Experimentally being tested on alfalfa, cole crops, onions, sorghum, sugar beets, sugarcane, and others. Being used outside the U.S. on rape, vegetables, and other crops.

IMPORTANT PESTS CONTROLLED: Alfalfa weevil, billbugs, chinch bugs, corn rootworm, Japanese beetle, root maggots, termites, thrips, webworms, white grubs, wireworms, and others.

APPLICATION: Used as a preplant or preemergence soil treatment. Apply to turf as a granular treatment when insects appear.

PRECAUTIONS: Do not feed treated grass clippings. Toxic to fish.

ADDITIONAL INFORMATION: Effective on soil-dwelling insects. Moderately toxic to fish and earthworms.

NAMES

PROPETAMPHOS, BLOTIC, DETMOL, OVIDIP, SAFROTIN, SERAPHOS

H_3C—O, S, H

P—O—C═C, O

H_5C_2—NH, C, CH_3

CH_3

O—CH

CH_3

(E)-0-2-Isopropoxy-carbonyl-1-methyl-vinyl-0-methyl ethylphosphoramidothioate

TYPE: Propetamphos is an organophosphorous insecticide, acting by contact and ingestion, with long residual activity.

ORIGIN: Sandoz Ltd. (Switzerland), 1969.

TOXICITY: LD_{50}-83 mg/kg.

FORMULATIONS: 50% EC, 1% liquid ready for use, dusts, 2% aerosol, 2% powders.

USES: Presently being sold for structural pest control purposes. May be used in food handling establishments.

IMPORTANT PESTS CONTROLLED: Adult mosquitoes, ants, bedbugs, black carpet beetle, body louse, cockroaches, crickets, fleas, houseflies, (resistant strains included), spiders, ticks, and others.

RATES: Use a rate of .5-1% ai.

APPLICATION: Apply where insects hide or congregate. Repeat as necessary.

PRECAUTIONS: Toxic to fish.

ADDITIONAL INFORMATION: Good residual activity, giving 12-15 weeks' control. No repellent activity.

NAMES

PHENAMIPHOS, BAY 68138, FENAMIPHOS, NEMACUR

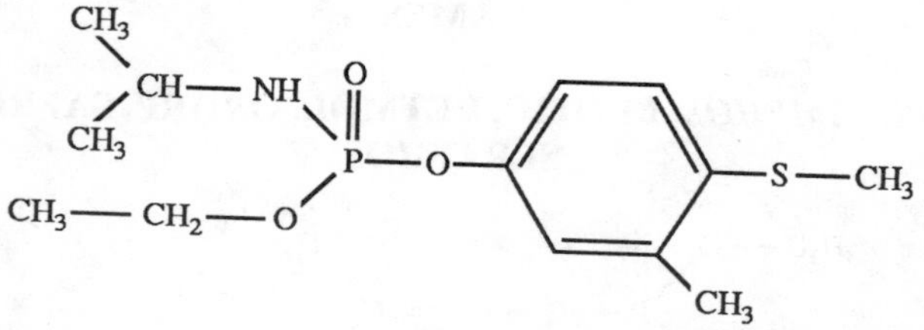

Ethyl-3-methyl-4-(methylthio) phenyl (1-methylethyl) phosphoramidate

TYPE: Nemacur is an organic-phosphate compound used as a contact and systemic nematocide-insecticide.

ORIGIN: Bayer AG of Germany, 1969. Being developed in the U.S. by Mobay Chem. Corp.

TOXICITY: LD_{50}-15 mg/kg.

FORMULATIONS: 3 EC, 5, 10, and 15% granules, 400 g/1 EC.

PHYTOTOXICITY: Alfalfa, squash, tomatoes, and certain ornamentals have shown some degree of injury when applied to the foliage. Soil applications are not normally phytotoxic.

IMPORTANT PESTS CONTROLLED: Nematodes and mole crickets. In the U.S., good results were obtained on alfalfa weevil, mites, fleahoppers, mealy bugs, aphids, thrips, and others.

USES: Apples, asparagus, bananas, Brussels sprouts, cabbage, caneberries, cherries, citrus, cocoa, cotton, garlic, grapes, okra, peaches, peanuts, pineapples, soybeans, strawberries, peppers, kiwifruit and turf. Used on field crops, tobacco, citrus, pineapple, peanuts, bananas, vegetables, and ornamentals in many countries.

RATES: Applied at 4-30 lb a.i./A.

APPLICATION: Being used as a soil treatment with or without incorporation, as a bare-root dip, a seed treatment, and as a foliar application. On turf irrigate within 6 hours of application.

PRECAUTIONS: Do not mix with pesticides which are alkaline in reaction. Weak on soil insects. Toxic to fish. Do not graze the treated area. Do not treat newly seeded turf.

ADDITIONAL INFORMATION: Not considered a fumigant. Nematodes must come into contact with this material to be controlled. Gives a long residual control. Tests show this material to be systemically active against a number of foliage-feeding insects. Absorbed by the roots.

NAME

HEXYTHIAZOX, ACORIT, SALIBRE, CALIBRE, DPX-Y 5893, MATACAR, SAVEY, STOPPER, TREVI, ZELDOX

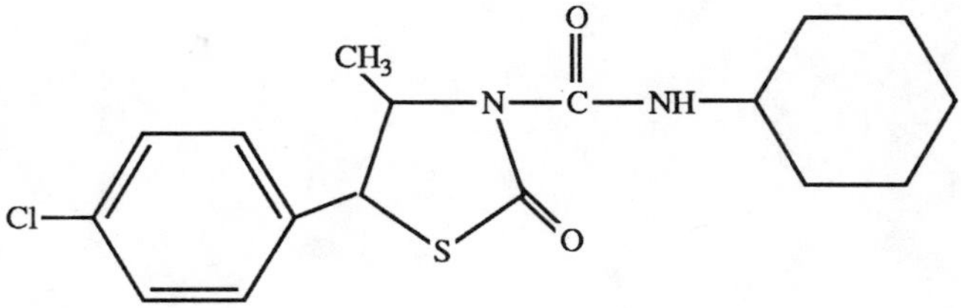

CH_3 N C O NH Cl S O

TYPE: DPX-Y 5893 is an carboxamide compound used as an ovicidal, and larvacidal acaricide.

ORIGIN: DuPont Chemical Co. and Nippon Soda., 1983. Sold outside the U.S. by a number of companies.

TOXICITY: LD_{50}-5000 mg/hg. May cause eye irritation.

FORMULATION: 50% WP.

PHYTOTOXICITY: Non-phytotoxic when used as directed.

USES: Experimentally being tested on apples, pears, grapes, almonds, citrus, stone fruit, strawberries, and ornamentals.

IMPORTANT INSECTS CONTROLLED: Mites.

RATES: Applied at .5-1 oz a.i./100 gal of water or at 1.25-2.5 oz. a.i./Acre.

APPLICATION: Apply when eggs are present prior to the adult infestation.

PRECAUTIONS: To be used on an experimental basis only. Toxic to fish. Do not mix with synthetic pyrethroid insecticides or diazinon, Trithron or Supracide.

ADDITIONAL INFORMATION: Adult mites are not controlled. Good residual activity of up to 50-60 days. Controls upon direct contact with the spray or from contact with treated plant surface. No systemic activity. Good ovicide activity. Compatible with other pesticides.

ORGANIC PHOSPHATES
Thiophosphoric Acid Prototypes

NAMES

PARATHION, ALKRON, ALLERON, COROTHION, EKATOX, ETILON, FOLIDOL, FOSFERNO, LETHALAIRE G-54, NIRAN, PANTHION, PARAMAR, PARATHENE, PHOSKIL, SNP, SOPRATHION, STRATHION, THIOPHOS

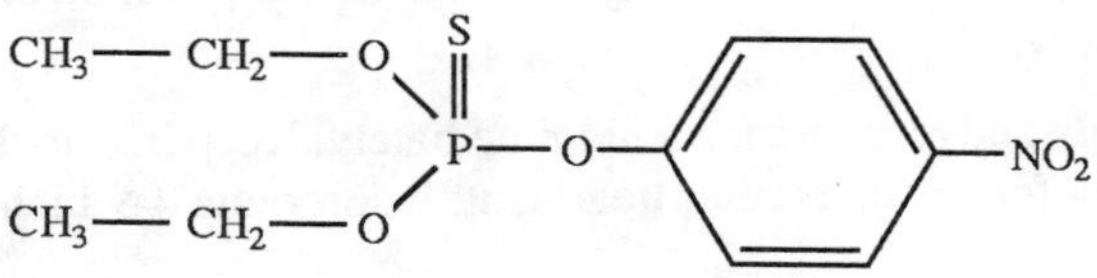

$$(CH_3-CH_2-O)_2P(=S)-O-C_6H_4-NO_2$$

O,O-Diethyl-O-4-nitrophenyl phosphorothioate

TYPE: Parathion is an organic phosphate insecticide-acaricide with both contact and stomach-poison activity.

ORIGIN: Bayer AG in Germany, 1947. Cheminova is the principle basic producer.

TOXICITY: LD_{50}-3 mg/kg. Readily absorbed through the skin.

FORMULATIONS: 15 and 25% WP, 2, 4, 6, and 8 EC, 1/2, 1, and 2% dusts, 10% granules, 10% aerosols. Sometimes mixed with other pesticides.

PHYTOTOXICITY: Injury has been reported on Bosc pears, cucurbits, certain ornamentals, and the apple varieties of McIntosh, Golden Delicious, R.I. Greening, Snow, Jonathan, and Duchess. Some hybrid sorghum varieties have been injured.

USES: Alfalfa, almonds, apples, apricots, artichokes, avocadoes, barley, beans, beets, blackeye peas, blueberries, caneberries, broccoli, Brussels sprouts, cabbage, carrots, cauliflower, celery, cherries, citrus, clover, collards, corn, cotton, cowpeas, cranberries, cucumbers, currants, dates, dewberries, eggplant, endive, figs, filberts, garlic, gooseberries, grapes, grasses, guavas, hops, kale, kohlrabi, lettuce, mangoes, melons, mustard greens, nectarines, oats, okra, olives, onions, peas, peaches, peanuts, pears, pecans, peppers, pineapples, plums, prunes, potatoes, pumpkins, quinces, radishes, rice, rutabagas, safflower, sweet potatoes, Swiss chard, tobacco, sorghum, soybeans, spinach, squash, strawberries, sugar beets, tomatoes, turnips, vetch, walnuts, wheat, and agricultural premises.

IMPORTANT PESTS CONTROLLED: Codling moth, scales, aphids, mealybugs, leafhoppers, mosquitoes, grasshoppers, mites, thrips, leaf miners, plum curculio, crickets, pear psylla, spittlebugs, armyworms, corn borers, corn earworms, and many others.

RATES: Applied at .1-1 lb actual/A.

APPLICATION:

1. Greenhouses-Use metal, pressure-packed aerosols. Aerate at least 1 hour before re-entering.
2. Foliage-Apply at a uniform rate with common application equipment. Repeat as necessary.
3. Soil-Apply and disc into the soil immediately. Keep livestock and persons out of the area for 48 hours after treatment. Incorporate 4-9 inches deep.

PRECAUTIONS: Do not apply while crops are in bloom, to avoid injury to bees. Incompatible with alkaline materials.

ADDITIONAL INFORMATION: Compatible with other insecticides and fungicides. Shows some fumigant action. It has the odor of garlic. Noncumulative in mammals. Does not accumulate in the soil. Used by public health authorities for mosquito control.

NAMES

***METHYLPARATHION*, BLADAN-M, FOLIDOL-M, METACIDE, METAFOS, METHYL NIRAN, METRON, NITROX, PARTRON-M, PENNCAP-M, TEKWAISA, WOFATOX**

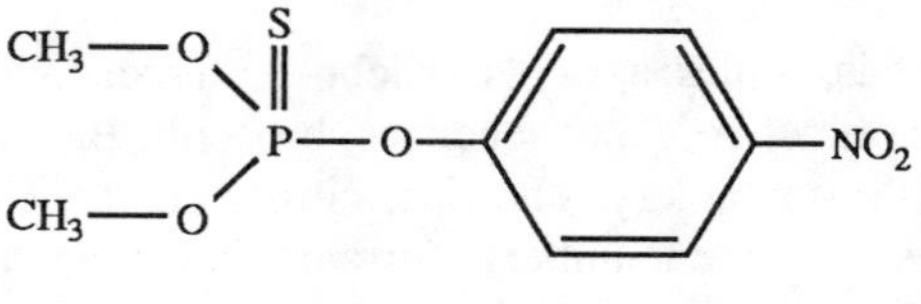

O,O-Dimethyl-o-p-nitrophenyl phosphorothioate

TYPE: Methyl Parathion is an organic phosphate insecticide-acaricide effective as a stomach and contact poison.

ORIGIN: Bayer AG in Germany, 1949. Cheminova is the basic producer.

TOXICITY: LD_{50}-9 mg/kg. Absorbed readily through the skin.

FORMULATIONS: 2, 4, 6, and 8 EC, 20-40% WP, 80% solution, 2.5 and 5% dusts. Sometimes sold mixed with other compounds.

PHYTOTOXICITY: Non-phytotoxic when used at the recommended dosages. Some injury has been reported on alfalfa and sorghum.

USES: Alfalfa, apples, almonds, blackeye peas, pumpkins, squash, eggplant, melons, nectarines, apricots, artichokes, barley, beans, beets, Brussels sprouts, broccoli, cabbage, carrots, cauliflower, celery, cherries, clover, collards, corn, cotton, cucumbers, gooseberries, grapes, grass, hops, kale, kohlrabi, lettuce, mustard, oats, onions, peaches, peanuts, pears, peas, peppers, plums, potatoes, prunes, rice, rutabagas, rye, safflower, sorghum, soybeans, spinach, strawberries, sugar beets, sunflowers, sweet potatoes, tobacco, tomatoes, turnips, vetch, wheat, pastures, ornamentals, and for general mosquito control. Use in pine forests for insect control.

IMPORTANT PESTS CONTROLLED: Aphids, armyworms, flea beetles, leafhoppers, leaf miners, scale, mealybugs, mites, boll weevils, thrips, and many others. Especially effective on boll weevils.

RATES: Applied at 1/4-2 lb actual/A.

APPLICATION: Apply evenly to foliage at a uniform rate. Repeat as necessary. Agitate while spraying.

PRECAUTIONS: Unprotected workers should not enter fields for at least 48 hours following treatment. Do not allow drift in areas where there are unprotected humans or animals. Destroy the empty container. Noncompatible with alkaline compounds. Toxic to bees. Toxic to fish and wildlife, so do not use where shrimp and crab are an important resource.

ADDITIONAL INFORMATION: Fast acting. Does not persist in the soil. No harmful effects have been noted on soil microorganisms. Used on irrigated pastures as a mosquito larvicide. Widely used in mosquito abatement programs.

NAMES

FENITROTHION, ACCOTHION, AGROTHION, BAY 41831, CYFEN, CYTEL, DICOFEN, FENSTAN, FOLITHION, MEP, METATHION, NOVATHION, NUVANOL, PESTROY, SUMANONE, SUMITHION, VERTHION

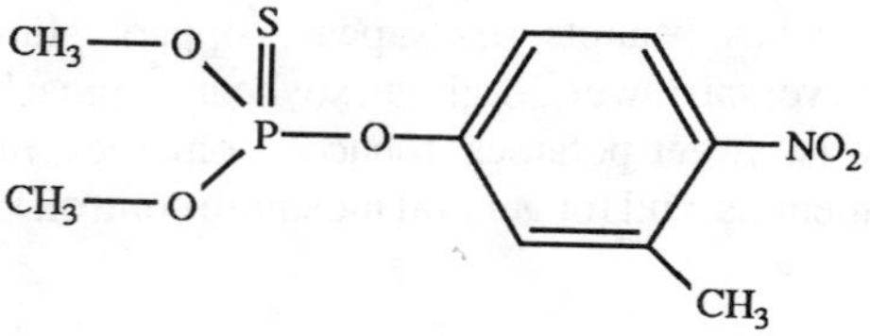

CH_3—O, S, P—O, NO_2, CH_3—O, CH_3

O,O-Dimethyl O-(3-methyl-4-nitrophenyl) phosphorothioate

TYPE: Fenitrothion is an organic phosphate insecticide killing as a stomach-poison and by contact action.

ORIGIN: Bayer AG in Germany, American Cyanamid Co. and Sumitomo Chemical Company, Ltd. of Japan, 1959.

TOXICITY: LD_{50}-330 mg/kg. May cause mild skin and eye irritation.

FORMULATIONS: 10, 50, 60, & 80% EC, 2-3% dusts, 40% WP, 3% granules, 1000 g/1 EC, 5% fog.

PHYTOTOXICITY: Cotton, Brassica crops, and certain fruit crops have been injured by high rates. Certain apple varieties may be russeted.

USES: Forest trees, indoor and outdoor mosquito control, and to control cockroaches. Being used outside the U.S. on tobacco, rice, mulberry, rape, orchards, grapes, forests, pastures, tea, cocoa, sugar beets, citrus, field and vegetable crops, cotton, cereals, coffee, soybeans, and sugarcane. Used in Europe and other countries as a public health insecticide, and as a stored product insecticide.

IMPORTANT PESTS CONTROLLED: Aphids, leafhoppers, plant hoppers, rice borers, mites, armyworms, bollworms, boll weevils, cockroaches, bed bugs, mosquitoes, flies, gnats, whiteflies, scale, thrips, codling moths, mealybugs, pear psylla, stem borers, spruce budworm, lygus, and many others.

RATES: Usually used at a concentration of .05-.075% active or at 1/2-1-1/2 lb actual/A.

APPLICATION: Thorough coverage is necessary. Apply when insects first appear. Repeat as necessary.

PRECAUTIONS: Harmful to bees, so do not spray on a flowering crop. Spray immediately if mixed with alkaline compound. Somewhat toxic to fish. May cause a slight yellowish stain on light colored surfaces.

ADDITIONAL INFORMATION: Gives a fast clean up with a long residual effectiveness. Somewhat effective against mites. Most effective against sucking and biting insects. Good penetrative action. Compatible with most other pesticides. Expresses ovicidal activity. Nonsystemic. Considered as effective as parathion, but much safer to handle. Long lasting activity.

NAMES

EPN, EPN (PIN), SANTOX

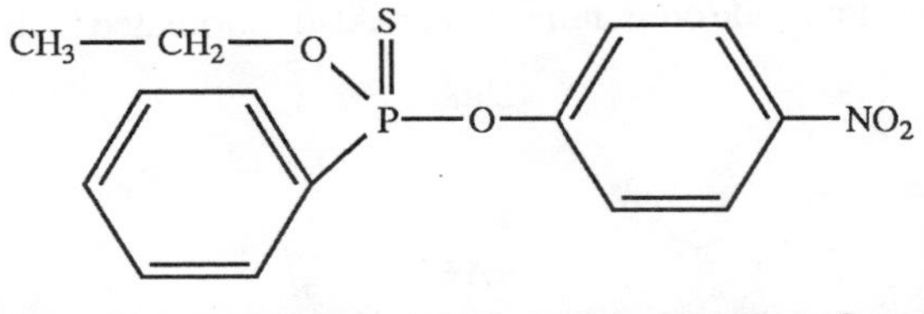

O-Ethyl O-p-nitrophenyl thionobenzenephosphonate

TYPE: EPN is an organic phosphate insecticide-acaricide that gives a high initial kill, as well as relatively long residual effectiveness.

ORIGIN: DuPont Chemical Company, 1949. Nisson Industries of Japan has also developed this compound and is the principle producer. No longer sold in the U.S.

TOXICITY: LD_{50}-14 mg/kg. Absorbed through the skin.

FORMULATIONS: 2 EC, 15% granules, 4 EC, 1.5% dust.

PHYTOTOXICITY: Injury has been reported on McIntosh and related varieties of apples. Otherwise, it is considered Non-phytotoxic.

USES: Almonds, apples, apricots, beans, beets, cherries, citrus, corn, cotton, grapes, lettuce, nectarines, olives, peaches, pears, pecans, plums, prunes, soybeans, sugar beets, tomatoes, walnuts, and outdoor mosquito larvae control. Used to a large extent in Japan on rice, vegetables, and other crops.

IMPORTANT PESTS CONTROLLED: Rice stem borer, boll weevils, Oriental fruit moth, cotton bollworms, plum curculio, fruit moths, peach tree borers, mites, European corn borers, scale, bud moths, leaf rollers, codling moths, pear psylla, aphids, thrips, armyworms, leaf miners, Mexican bean beetles, and many others.

RATES: Applied at 1/8-12 lb actual/A or 1-3 actual/100 gal of water.

APPLICATION: Thoroughly cover the entire plant surface. Begin with the first signs of infestation and repeat as necessary.

PRECAUTIONS: Do not use with Bordeaux or zinc sulfate-lime sprays. Toxic to fish. Toxic to bees. Do not use within 10 days of a DCPA herbicide application. No longer sold in the U.S.

ADDITIONAL INFORMATION: Acts similar to Parathion, only it is somewhat more persistent. Compatible with most other pesticides. Used to some extent as a mosquito larvicide on standing water. Also formulated with methyl parathion.

NAMES

***DIAZINON,* BASUDIN, DAZZEL, DIAZIDE, DIAZITAL, DIAZOL, GARDENTOX, KAYAZINON, KAYAZOL, KNOX-OUT, NEDCIDOL, NIPSAN, NUCIDOL, SAROLEX, SPECTRACIDE**

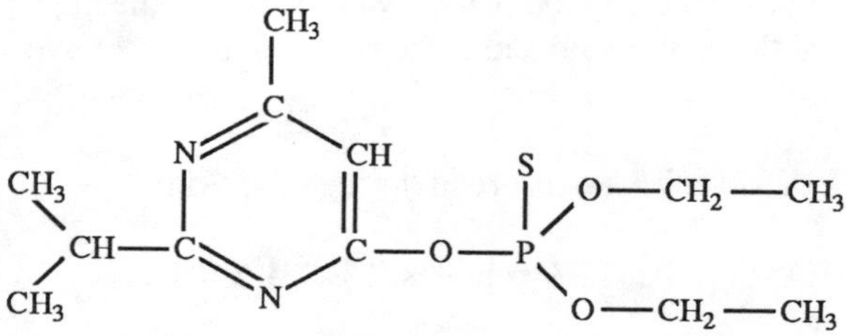

O-O-Diethyl-O-(2-isopropyl-6-methyl-5-pyrimidinyl) phosphorothioate

TYPE: Diazinon is an organic phosphate insecticide-acaricide with contact and stomach-poison activity.

ORIGIN: CIBA-Geigy Corporation, 1956. Produced by a number of manufacturers today.

TOXICITY: LD_{50}-300 mg/kg. May be absorbed through the skin.

FORMULATIONS: 40% and 50% WP, 4 EC, 60% EC, dusts, 14% granules, .5% aerosols, 4 and 6 lb/gal oil solution. Also sold mixed with fertilizer.

PHYTOTOXICITY: Non-phytotoxic, except to Stephanotis and African Violets. Russeting on green and yellow varieties of apples has occurred. Some lettuce varieties are injured. Some ornamentals are injured.

USES: Alfalfa, almonds, apples, apricots, bananas, barley, beans, Bermuda grass, beets, blueberries, broccoli, Brussels sprouts, cabbage, caneberries, carrots, cauliflower, celery, cherries, citrus, clover, collards, corn, cotton, cowpeas, cranberries, cucumbers, dandelion, endive, figs, filberts, grapes, grass hay, hops, kale, lespedeza, guar, lettuce, lima beans, melons, mushroom houses, mustard, nectarines, oats, olives, onions, parsley, parsnips, peaches, peanuts, pears, peas, pecans, peppers, pineapples, potatoes, plums, prunes, radishes, rye, sorghum, soybeans, spinach, squash, strawberries, sugar beets, sugarcane, sweet potatoes, Swiss chard, tobacco, tomatoes, trefoil, turnips, watermelons, watercress, walnuts, wheat, ornamentals, pastures, sheep, turf, and agricultural premises. Also used on rice outside of the U.S.

IMPORTANT PESTS CONTROLLED: Cockroaches, mites, codling moths, aphids, scale, leafhoppers, pear psylla, corn earworms, Drosophila, lygus bugs, houseflies, ants, ticks, silverfish, fleas, chiggers, lice, mosquitoes, and many others.

RATES: Applied at 1/4-1 lb a.i./100 gal of water or 1/4-2 lb actual/A.

APPLICATION: Apply with common application equipment at a uniform rate on fruits and vegetables. Repeat as necessary. Livestock barns are treated for houseflies, giving a residual control for up to 8 weeks. (Livestock should be removed for at least 4 hours after treatment.) Sheep can be dipped in a dilute solution for fly, lice, tick, and flea control. May be applied to the soil as a preplant insecticide. Apply to sheep as a spray at least 14 days prior to slaughter. Used as a crack and crevice treatment.

PRECAUTIONS: Ducks and geese are highly susceptible to this compound. Do not mix with copper compounds. Toxic to bees. Do not use on sod farms or golf courses in the U.S.

ADDITIONAL INFORMATION: Long residual effects. No off-flavors have resulted on harvested fruits and vegetables. Compatible with other pesticides. Used as a seed treatment on certain crops.

NAMES

ISAZOPHOS, CGA-12223, MIRAL, TRIUMPH, VICTOR

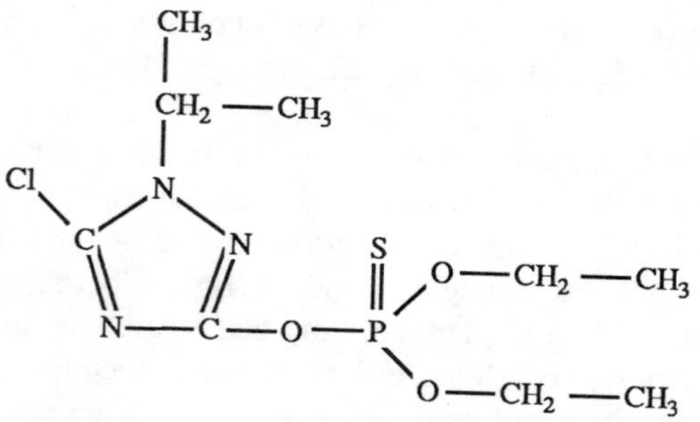

O-(5-Chloro-1-methylethyl) 1H-1,2,4-triazol-3-yl) O,O-diethyl phosphorothioate

TYPE: MIRAL is an organic phosphate compound used as a contact and stomach-poison insecticide-nemanticides.

ORIGIN: CIBA-Geigy. Ltd., 1973.

TOXICITY: LD_{50}-60 mg/kg.

FORMULATIONS: 4 EC, 2 EC, 1% and 2% granules, 1 EC.

PHYTOTOXICITY: Do not use on tobacco or potatoes.

IMPORTANT PESTS CONTROLLED: Chinch bugs, corn rootworms, cutworms, Japanese beetle, leaf hoppers, mole crickets, nematodes, stem borers, sweet corn, maggots, webworms, white grubs, and other soil insects.

USES: Turf. Used on bananas, corn, cotton, rice, sugar beets, turf, vegetables, and others outside the U.S.

RATES: Applied at .1-2 lb. a.i./Acre.

APPLICATION: Applied as a soil insecticide. Incorporate into the soil 3-6 inches deep or applied as a banded application. On turf apply broadcast and let rainfall or irrigation carry it into he soil.

PRECAUTIONS: Toxic to fish.

ADDITIONAL INFORMATION: Applied to the soil, it is systemic against foliar insects, as well as soil insects. For nematode control, it must come in contact with the nematodes. Gives up to 2 months control of turf insects.

NAMES

TRIAZOPHOS, HOSTATHION

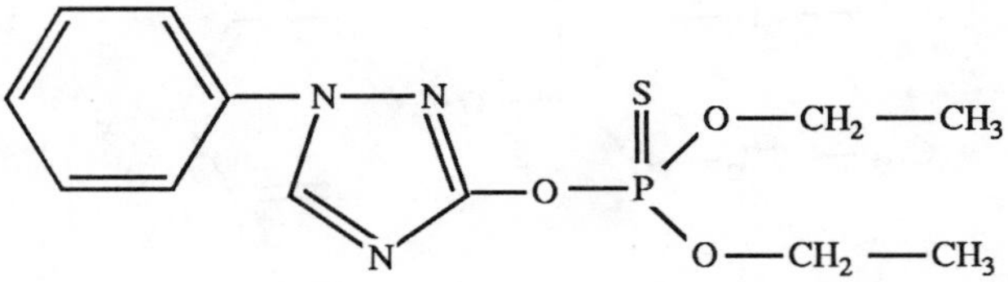

1-Phenyl-3-(O,O-diethyl-thionophosphate)-1,2,4-triazole

TYPE: Hostathion is a phosphoric-acid ester compound used as a contact and stomach-poison insecticide-acaricide.

ORIGIN: Hoechst AG of Germany, 1971.

TOXICITY: LD_{50}-64 mg/kg.

FORMULATIONS: 40% EC, granules, seed dressings, 25 and 40% ULV, 30% WP.

USES: Being used outside the U.S. on cotton, vegetables, cereals, coffee, potatoes, corn, fruit, palm, rice, citrus, and others.

RATES: Applied at .02-.1% a.i.

IMPORTANT PESTS CONTROLLED: Thrips, aphids, bollworms, nematodes, caterpillars, lygus, psylla, leafhoppers, stem borers, cutworms, armyworms, mites, and many others.

APPLICATION: Apply when insects appear, and repeat as necessary.

PRECAUTIONS: Do not use on any food or feed crop until registration has been granted. Toxic to bees and fish.

ADDITIONAL INFORMATION: Possesses insecticidal, acaricidal, and nematocidal activity against leaf nematodes. Good initial kill, with residual control. Nonsystemic and broad-spectrum in activity, with the ability to penetrate leaf tissue.

NAMES

***PYRIDAPHENTHION*, OFNACK, OFNAK, OFUNACK**

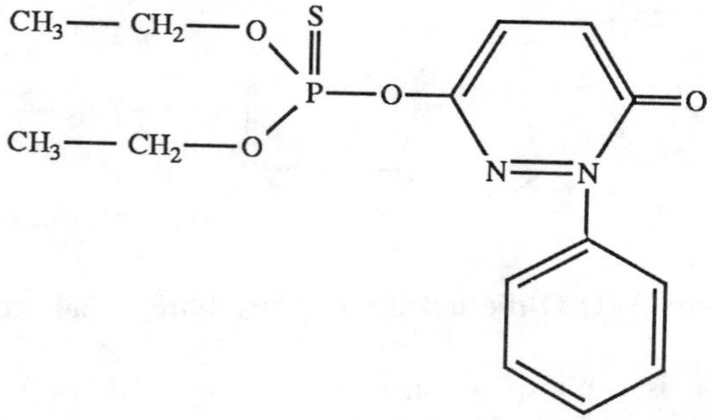

O, O-Dimethyl-O-(3-oxo-2-phenyl-dihydro-pyridazine-6-yl) phosphorothionate

TYPE: Ofnak is an organic-phosphate insecticide that kills both by contact and by ingestion.

ORIGIN: Mitsui Toatsu Chemicals Inc. of Japan, 1971.

TOXICITY: LD_{50}-769 mg/kg.

FORMULATIONS: 40% EC, 50% WP, dusts.

PHYTOTOXICITY: Non-phytotoxic when used as directed.

USES: Outside the U.S. on rice, fruit trees, vegetables, and cereals, and in public health programs.

IMPORTANT PESTS CONTROLLED: Rice stem borer, green rice leafhopper, plant hopper, phytophagous mites, and many other sucking and chewing insects.

RATES: Apply at 60-100 g a.i./A.

APPLICATION: Apply when insects appear and repeat as necessary.

ADDITIONAL INFORMATION: Relatively Non-toxic to fish.

NAMES

ISOXATHION, KARPHOS

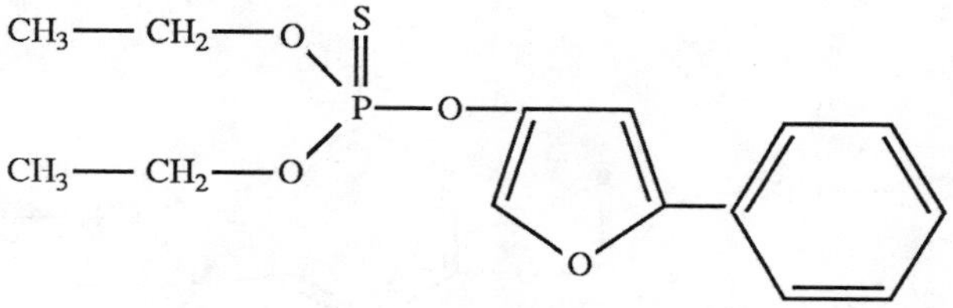

O,O-Diethyl O-(5-phenyl 3-isoxazolyl) phosphorothioate

TYPE: Karphos is an organic phosphate compound used as a contact and stomach-poison insecticide.

ORIGIN: Sankyo Co. Ltd. of Japan, 1972.

TOXICITY: LD_{50}-112 mg/kg.

FORMULATIONS: 50% EC, 40% WP, 3% granule, 2 and 3% dusts.

IMPORTANT PESTS CONTROLLED: Scales, armyworms, diamondback moth, cutworms, grasshoppers, aphids, borers, mealybugs, hoppers, gall midge, caterpillars, beetles, mites, and many others.

USES: Used in Japan on citrus, vegetables, orchards, forage crops, ornamentals, and rice.

RATES: Applied at 330-500 ppm.

APPLICATION: Apply when insects appear and repeat as necessary. Also used as a soil insecticide.

PRECAUTIONS: Not for sale or use in the U.S. Toxic to fish. Do not mix with alkaline pesticides.

ADDITIONAL INFORMATION: Considered to be a broad-spectrum insecticide. No systemic activity and no fumigant activity.

NAMES

***PIRIMIPHOS-ETHYL*, PRIMICID, PP211, FERNEX, PRIMOTEC, SOLGARD**

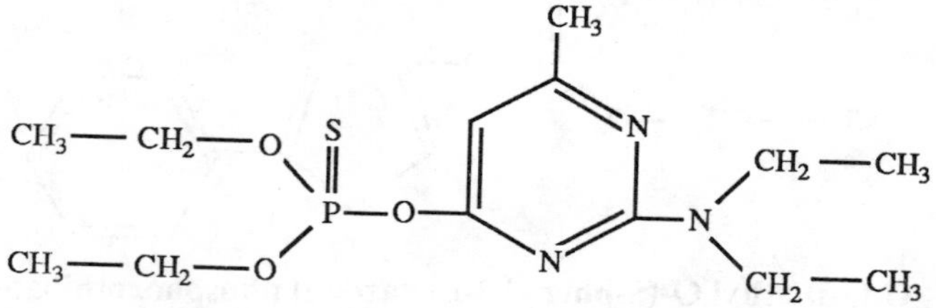

$$
\begin{array}{c}
CH_3-CH_2-O \quad S \\
\quad \backslash \;\; \| \\
\quad P-O-(\text{pyrimidine ring: } 6\text{-}CH_3,\ 2\text{-}N(CH_2-CH_3)_2) \\
\quad / \\
CH_3-CH_2-O
\end{array}
$$

O,O-Diethyl O-(2-(diethylamino) 6-methyl-4 pyrimidnyl) phosphorothioate

TYPE: Pirimiphos-ethyl is a pyrimidine phosphate compound used as a broad-spectrum insecticide, effective by contact and fumigant action.

ORIGIN: ICI of England, 1971.

TOXICITY: LD_{50}-138 mg/kg. May cause eye irritation.

FORMULATIONS: 20% WP seed dressing, 5 and 10% granules, 4 EC.

PHYTOTOXICITY: Slight phytotoxicity has been noted with high rates of seed dressing. Non-phytotoxicity to turf grasses.

IMPORTANT PESTS CONTROLLED: Armyworms, seed corn beetles, banana borers, wireworms, corn rootworm, onion fly, carrot fly, scale, sod webworms, Japanese beetle, chinch bugs, rust mite, caterpillars, and many others.

USES: Turf, vegetable, fruit, field, and ornamental crops outside the U.S.

RATES: Applied at 1-1.5 lb a.i./A.

APPLICATION: Applied as a foliar spray, root dip, soil drench, or seed dressing.

PRECAUTIONS: Not for sale in the U.S. Toxic to bees. Corrosive to iron.

ADDITIONAL INFORMATION: Effective against both soil and foliage insects. Kills upon contact. A wide-spectrum insecticide. Does not harm earthworms.

NAMES

***PIRIMIPHOS-METHYL*, ACTELLIC, BLEX, SILOSAN**

S O—CH_3

CH_3— —O—P

O—CH_3

N N

CH_3— CH_2— N—CH_2—CH_3

2-Diethylamino-6-methylpyrimidin-4-yl dimethyl phosphorothionate

TYPE: Pirimiphos is an organic phosphate compound used as a broad-spectrum contact insecticide.

ORIGIN: ICI of England, 1970. Being sold in the U.S. by ICI Americas.

TOXICITY: LD_{50}-2050 mg/kg.

FORMULATIONS: 25% and 50% EC, 50 ULV, 20% liquid seed treatment, 2% dust, 2% aerosol.

PHYTOTOXICITY: Non-phytotoxic when used as directed.

IMPORTANT PESTS CONTROLLED: Store-grain pests, cockroaches, mosquitoes, lice, fleas, bedbugs, houseflies, ants, and many fruits and vegetable pests

USES: Corn, sorghum, kiwi fruit, wheat, and rice in the U.S. Used in Europe on stored-grain insects and in public health. Crop-protection usages on rice, citrus, cole crops, sugar beets, grapes, lettuce, olives, tomatoes, and others are being utilized in certain countries.

APPLICATION:

1. Public health-Apply where insects hide or congregate.
2. Stored grains-Apply to the bin walls, as a spray on bagged produce, or by mixing with the grain.
3. Crop protection-Apply when insects appear, and repeat as necessary.

PRECAUTIONS: Toxic to fish.

ADDITIONAL INFORMATION: Appears to be more effective at higher temperature than most store-grain insecticides. Persists on surfaces longer than most insecticides. Fast acting. Persistent on inert surfaces, such as concrete or brick. Has fumigant activity.

NAMES

ETRIMPHOS, ETRIMFOS, EKAMET, SATISFAR

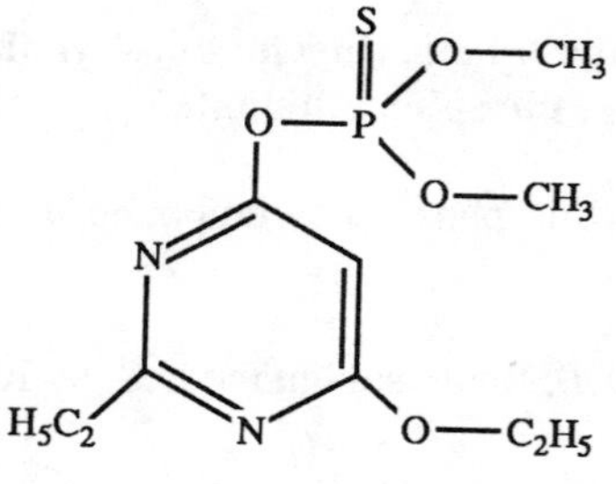

O-6-Ethoxy-2-ethyl-pyrimidin-4-yl-O,O-dimethyl-phosphorothioate

TYPE: Etrimfos is a pyrimidinyl-thionophosphate insecticide of low toxicity with both contact and stomach activity.

ORIGIN: Sandoz Ltd. (Basle, Switzerland), 1971.

TOXICITY: LD_{50}-1600 mg/kg.

FORMULATIONS: 50% EC, 5% G, 400 g/1ULV, 30 and 50 g/1LS, 50% DP, 2% DP.

PHYTOTOXICITY: Injury has been reported on cherries and certain apple varieties with the EC.

USES: Alfalfa, citrus, crucifers, fruit trees, grapes, maize, olives, potatoes, rice, tobacco, and vegetables outside of the U.S. Also used to protect stored products, such as grains, oilseeds, peanuts, beans, soybeans, potatoes, tobacco, coffee, cocoa, processed foods, and others.

IMPORTANT PESTS CONTROLLED: Alfalfa weevils, aphids, beetles, cabbage moth, carrot fly, Colorado beetle, diamondback moth, European corn borer, grapeberry moth, lepidopterous pests, midges, rice stemborers, and others.

RATES: Applied at .25-.75 kg a.i./ha. For stored products, apply at 2-7 ppm ai.

APPLICATION: Apply when insects appear, and repeat as necessary.

PRECAUTIONS: Not for sale or use in the U.S. Toxic to fish.

ADDITIONAL INFORMATION: Compatible with most other insecticides and fungicides, with the exception of strongly alkaline ones. Activity lasts for 7-11 days. Gives stored products protection for a year or more.

NAMES

DEMETON, DEMOX, MERCAPTOPHOS, SYSTEMOX, SYSTOX

$(CH_3-CH_2-O)_2P(=S)-O-CH_2-CH_2-S-CH_2-CH_3$ I

$(CH_3-CH_2-O)_2P(=O)-S-CH_2-CH_2-S-CH_2-CH_3$ II

O, O-Diethyl O-2-(ethylthio) ethyl phosphorothioate mixture with O,O-Diethyl S-2-(ethylthio) ethyl phosphorothioate

TYPE: Systox (Demeton) is an organic phosphate insecticide-acaricide with contact and systemic activity.

ORIGIN: Bayer AG in Germany, 1950. Licensed to be sold in the U.S. by Mobay Chemical Corp. The first systemic insecticide to be approved on food crops. No longer sold in the U.S.

TOXICITY: LD_{50}-2.5 mg/kg. Readily absorbed through the skin.

FORMULATIONS: EC 2 and 6 lb a.i./gal.

PHYTOTOXICITY: McIntosh and Golden Delicious apples, as well as pears, have shown injury. Injury has occurred on African violets, Cibotium ferns, and Croft lilies.

USES: Outside the U.S. Alfalfa, almonds, apples, apricots, barley, beans, blackberries, broccoli, Brussels sprouts, cabbage, cauliflower, celery, clover, cotton, dewberries, eggplant, filberts, gooseberries, grapefruit, grapes, hops, lemons, lettuce, loganberries, muskmelons, nectarines, oats, oranges, peaches, pears, peas, pecans, peppers, pineapples, plums, potatoes, prunes, raspberries, sorghum, strawberries, sugar beets, tomatoes, walnuts, wheat, and ornamentals.

IMPORTANT PESTS CONTROLLED: Aphids, leafhoppers, whiteflies, thrips, mites, and many others.

RATES: Applied at 1/8-3/4 lb actual/A.

APPLICATION: Penetrates into, and is translocated in, the sap stream of plants, from either foliage or roots, if applied as a soil drench. Sometimes applied as a seed treatment before planting.

PRECAUTIONS: Wildlife in treated areas may be killed. Do not use on greenhouse crops. Toxic to fish. Incompatible with Bordeaux, lime sulfur, and lime. Do not enter treated areas until spray residues have dissipated. Compatibility is doubtful with organic mercury, zinc arsenate, cryolite, calcium arsenate, Paris Green, and Cyprex. Do not let granules come into immediate contact with the seed. Do not graze treated fields for 21 days after application. Toxic to bees. Treated nursery stock should not be handled within 5 days of application. No longer sold in the U.S.

ADDITIONAL INFORMATION: Not effective against chewing insects, but as a systemic it is effective against sucking insects, making it a selective insecticide with long residual action. No off-flavor has resulted in harvested crops.

NAMES

OXYDEMETON-METHYL, METASYSTEMOX, METASYSTOX-R, METILMERCAPTOFOSOKSID, MSR, PHOSPHOROTHIOATE

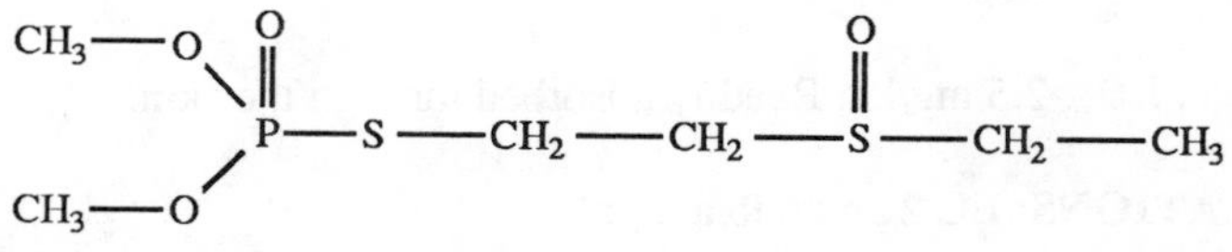

$$(CH_3-O)_2P(=O)-S-CH_2-CH_2-S(=O)-CH_2-CH_3$$

S-((2-Ethylsulfinyl)ethyl) O,O-dimethyl phosphorothioate

TYPE: Metasystox-R is a selective, systemic, and contact insecticide-acaricide.

ORIGIN: Developed by Bayer AG in Germany, 1960. Produced in the U.S. by Mobay Chemical Corp.

TOXICITY: LD_{50}-50 mg/kg. May cause eye irritation.

FORMULATIONS: 2 EC, 50% concentrate, 500 g/1 soluble concentrate.

PHYTOTOXICITY: Some injury to certain ornamentals has been reported, especially in combination with other pesticides.

USES: Alfalfa, apples, apricots, beans, broccoli, Brussels sprouts, cabbage, caneberries, cauliflower, cherries, citrus, clover, corn, cotton, eggplant, crab apples, cucumbers, filberts, lettuce, onions, mint, grapes, melons, nectarines, peaches, pears, plums, peas, safflower, potatoes, prunes, pumpkins, quinces, squash, sugar beets, peppers, turnips, watermelons, walnuts, Christmas trees, and ornamentals. Used outside the U.S. on these and other crops.

IMPORTANT PESTS CONTROLLED: Mites, aphids, whiteflies, sawflies, thrips, leafhoppers, and many others.

RATES: Applied at 1/4-1/2 lb actual/A.

APPLICATION: Apply to the foliage as a spray, or to roots as a soil drench. Capable of being translocated into the sap stream, giving it a residual effectiveness, since it is not washed off. May be injected into ornamental trees.

PRECAUTIONS: Do not handle treated ornamentals until the spray has dried. Incompatible with alkaline compounds. Harmful to bees. Toxic to fish and wildlife. Birds feeding on treated areas may be killed. Do not use in combination with EUPAREN or MORESTAN. Used in the U.S. on ornamentals only by injection or as a soil drench.

ADDITIONAL INFORMATION: Selective, giving quick knockdown effects. Compatible with most insecticides and fungicides, except those alkaline in nature. Injected into ornamental trees for systemic insect control. Widely used as a rose insecticide in the home and garden market. May be used on greenhouse plants. Widely used outside the U.S.

NAMES

MEVINPHOS, DURAPHOS, GESFID, APAVINPHOS, FINIPHOS, MEDVIDRIN, MENITE, PHOSDRIN, PHOSFENE

$$(CH_3-O)_2P(=O)-O-C(CH_3)=CH-C(=O)-O-CH_3$$

3-[(Dimethoxyphosphinyl)oxy]-2-butenoic acid methyl ester

TYPE: Phosdrin is an organic phosphate insecticide-acaricide with contact and systemic activity.

ORIGIN: Shell Chemical Company, 1953.

TOXICITY: LD_{50}-3.7 mg/kg. Readily absorbed through the skin, lungs, and mucous membranes.

FORMULATIONS: EC 2 to 4 lb a.i./gal, 10.3 lb active water-soluble solution, aerosols, and dusts.

PHYTOTOXICITY: No injury has been reported, and no harmful residue remains in the soil.

USES: Alfalfa, apples, artichokes, beans, beets, broccoli, Brussels sprouts, cabbage, cantaloupes, carrots, cauliflower, celery, cherries, citrus, clover, collards, corn, cucumbers, eggplant, grapes, kale, lettuce, melons, mustard greens, onions, okra, parsley, peaches, pears, peas, peppers, plums, potatoes, raspberries, sorghum, spinach, squash, strawberries, tomatoes, trefoil, turnips, watermelons, walnuts, and greenhouses. Used outside of the U.S. on these and many other crops.

IMPORTANT PESTS CONTROLLED: Aphids, leaf rollers, orange tortrix, weevils, armyworms, salt marsh caterpillar, loopers, corn earworms, cutworms, leaf miners, chinch bugs, grasshoppers, lygus bugs, mites, thrips, and many others.

RATES: Applied at 1/8-2 lb actual/A.

APPLICATION:

1. Foliage — Begin when insects first appear and repeat as often as necessary to obtain control. Apply uniformly.

2. Greenhouses — Apply aerosol formulations to closed greenhouses. Keep greenhouse closed tightly at least 2 hours after treating. Ventilate 1 hour before entering.

PRECAUTIONS: Do not contaminate streams or ponds. Corrosive to steel. Incompatible with alkaline compounds. Toxic to bees. Do not apply with hand-operated equipment. Toxic to birds and fish.

ADDITIONAL INFORMATION: Short residual activity. Compatible with insecticides and fungicides, except strongly alkaline materials. Does not accumulate in the soil. No adverse effects on soil microorganisms have been noted. Controls most insects within minutes, since it breaks down rapidly.

NAMES

QUINALPHOS, CHINALPHOS, BAY 77049, BAYRUSIL, EKALUX, KNAVE, SAVALL, TOMBEL

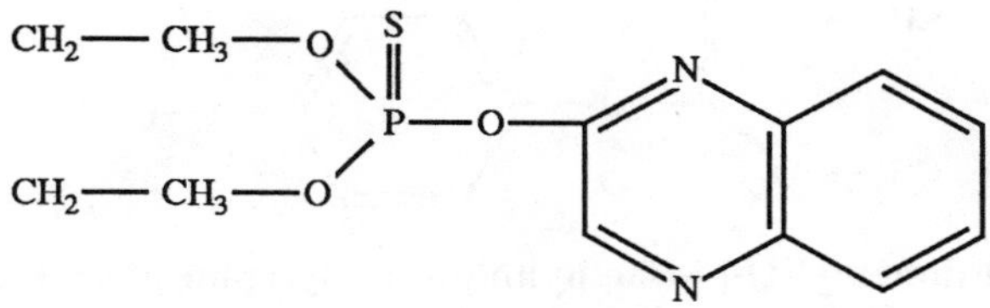

O,O-Diethyl-O-quinoxalin-2-yl-phosphorothioate

TYPE: Quinalphos is an organic phosphate insecticide used against sucking and biting insects as a contact and stomach-poison.

ORIGIN: Bayer AG of Germany and Sandoz Ltd. of Switzerland, 1969.

TOXICITY: LD_{50}-65 mg/kg.

FORMULATIONS: 25% EC, 5% granules, 200 & 250 g/1 EC, 5% dust.

PHYTOTOXICITY: Slight injury has been observed on certain fruit trees.

USES: Used outside the U.S. on beets, citrus, cotton, grapes, ornamentals, peanuts, potatoes, rice, tea, vegetables, and others.

IMPORTANT PESTS CONTROLLED: Aphids, bollworms, borers, caterpillars, diamondback moth, leafhoppers, mealybugs, mites, plant hoppers, scale, thrips, and others.

RATES: Used at a .025-.05% concentration of active ingredient.

APPLICATION: Apply when insects appear, and repeat as necessary.

PRECAUTIONS: Do not use in the U.S. Toxic to bees, so do not spray on open blossoms. Do not mix with alkaline compounds. Somewhat toxic to fish.

ADDITIONAL INFORMATION: A contact and stomach-poison with fast killing action. Compatible with most other pesticides. Use the spray on the same day it is mixed with water. Applications should be made every 10-14 days if insects reappear. Some activity against mites.

NAMES

***FENTHION,* BAYCID, BAYTEX, ENTEX, LEBAYCID, QUELETOX, SPOTTON, TIGUVON**

$$(CH_3O)_2P(=S){-}O{-}C_6H_3(CH_3){-}S{-}CH_3$$

O,O-dimethyl-O-[4-(methylthio)-m-tolyl] phosphorothioate

TYPE: Fenthion is an organic phosphate insecticide-acaricide with contact and stomach-poison activity and a long residual activity.

ORIGIN: Bayer AG in Germany, 1957. Sold in the U.S. by Mobay Chemical Corp.

TOXICITY: LD_{50}-250 mg/kg. More toxic to fowl than to mammals.

FORMULATIONS: 4 EC, 25, 40, and 50% WP, 3% dust, 2% granules, 1000 g/1 EC.

PHYTOTOXICITY: Considered Non-phytotoxic when used at the recommended rates. Do not apply to Hawthorne, American linden, sugar maple, or the rose variety Delightful. Some injury has occurred to certain varieties of apples and cotton.

USES: Registered for use as a mosquito larvicide, in households, on agricultural premises, alfalfa, rice, pasture grass, cattle, domestic pets, and on ornamentals. Used outside the U.S. for these, as well as on cotton, vegetables, cereals, fruit trees, corn, olives, rape, grapes, rice, citrus, tea, tobacco, sugarcane, and other crops.

IMPORTANT PESTS CONTROLLED: Flies, mosquitoes, roaches, ticks, lice, bedbugs, crickets, armyworms, cattle grubs, thrips, leaf miners, codling moth, psylla, bollworms, horn flies, fleas, aphids, leafhoppers, ants, mites, and others.

RATES: Applied at 1/2-1 lb actual/100 gal of water.

APPLICATION:

1. Foliage and premises-Apply evenly at a uniform rate. Repeat as necessary. Used for mosquito larvae control.
2. Livestock-Applied by a spray, pour-on, or backrubber. Used to paint barns, etc.

PRECAUTIONS: Do not spray plant foliage when temperatures exceed 90°F. Only trained personnel should use in households. Avoid excessive wetting of plastic, tile, rubber, etc. Do not spray flowering crop to protect bees. Noncompatible with highly alkaline pesticides. Do not apply as a space spray. Toxic to aquatic life. Toxic to bees. Do not apply to young, stressed, or sick animals. Do not apply for mosquito control in areas containing fish, shrimp, crabs, or crayfish.

ADDITIONAL INFORMATION: Compatible with other pesticides, except those which are highly alkaline. It has given control of insects in stored products from 4-16 months. It gave 100% control of mosquitoes after 42 weeks, when applied to the sides of barns. Sometimes mixed with paints. Tiguvon is the trade name when used on livestock. Used outside the U.S. to control birds when sold under the trade name Queletox. Used for treating walls made of a wide variety of materials. Does not stain colored surfaces. Gives residual effectiveness even on alkaline surfaces. Applied to pets by veterinarians as a subcutaneous injection for insect control. Widely used in mosquito-abatement programs. Used by pest-control operators in indoor and outdoor areas.

NAMES

***FENSULFOTHION*, DASANIT, DMSP, TERRACUR-P**

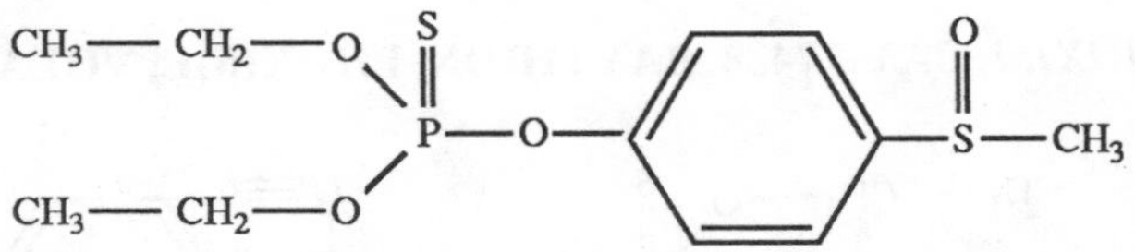

CH_3— CH_2—O, CH_3— CH_2—O, P(=S)—O—(phenylene)—S(=O)—CH_3

O,O-Diethyl O-(4-(methylsulfinyl)phenyl) phosphorothioate

TYPE: Dasanit is an organic phosphate compound used as a soil-contact insecticide-nematocide.

ORIGIN: Bayer AG of Germany, 1957. Produced and developed in the U.S. by Mobay Chemical Corporation.

TOXICITY: LD_{50}-4 mg/kg.

FORMULATIONS: 5, 10, and 15% granules, 6 EC.

PHYTOTOXICITY: Considered Non-phytotoxic if used according to label directions.

USES: Citrus, peanuts, corn, tobacco, sweet potatoes, tomatoes, sugar beets, bananas, pineapples, sorghum, soybeans, sugarcane, potatoes, onions, ornamentals, and turf. Used outside the U.S. on a number of other crops, such as pineapple, bananas, cotton, peanuts, vegetables, pastures, rice, and others.

IMPORTANT PESTS CONTROLLED: Corn rootworm, wireworm, nematodes, cabbage maggot, onion maggot, and others.

RATES: Applied at rates of 1/2-40 lb a.i./A, either band or broadcast.

APPLICATION: Applied to the soil at, or before, planting. On ornamentals, it can be used as a preplant treatment, working into the top 4-6 inches of soil, or mixed with the potting soil. Also effective as a soil drench.

PRECAUTIONS: Do not apply to plant foliage other than turf grasses. Do not treat food crops grown in greenhouses. Toxic to fish. Do not mix with alkaline compounds. Toxic to bees. Do not place treated band closer together than 6 inches.

ADDITIONAL INFORMATION: Kills primarily upon contact, so placement and proper mixing in the soil are essential. Protects plants for up to 4 months, and up to 9 months' protection of turf has been noted. May be applied with liquid fertilizer.

NAMES

***PHOXIM*, BAY-77488, BAYTHION, PHOXIME, VOLATON**

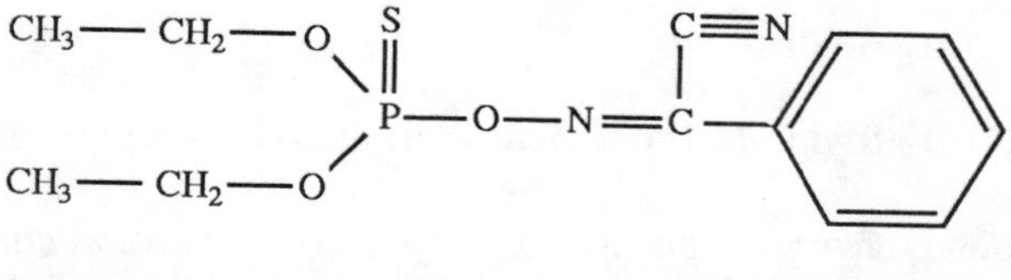

CH_3—CH_2—O, S, C≡N, P—O—N=C—, CH_3—CH_2—O

(Diethoxy-thiophosphoryloxyimino)-phenylacetonitrile

TYPE: Phoxim is an organic phosphate compound used as a contact and stomach-poison insecticide.

ORIGIN: Bayer AG of Germany, 1968. Being developed in the U.S. by Mobay Chemical Corp.

TOXICITY: LD_{50}-1680 mg/kg.

FORMULATIONS: 4 EC, 5% & 10% granules, 500 & 600 g/1 EC.

PHYTOTOXICITY: Injury has been reported on cotton.

USES: Being used on cotton, pastures, bananas, potatoes, tobacco, peanuts, vegetables, corn, and sugar beets outside of the U.S. Also used to control stored-product pests.

IMPORTANT PESTS CONTROLLED: Armyworms, diamondback moth, caterpillars, aphids, cabbage looper, bollworm, corn earworms, potato beetles, boll weevils, mosquitoes, houseflies, roaches, grasshoppers, mites, and onion maggot.

RATES: Applied 1/4-2 lb a.i./A.

APPLICATION: Apply as a foliar insecticide or as soil applications that are to be incorporated. As a soil insecticide, apply as a preplant treatment.

PRECAUTIONS: May cause yellow stains on light-colored surfaces. Do not mix with alkaline materials. Do not tank mix with wettable-powder pesticides. Crop injury has been caused when applied during high temperatures. Toxic to bees. Toxic to fish.

ADDITIONAL INFORMATION: A broad-spectrum insecticide with low mammalian toxicity. Gives good residual control. Fast knockdown effects. Ultra-low volume foliar sprays have proven successful. Most effective for the control of Lepidoptera and Coleoptera insects. Compatible with other pesticides. Also used to control stored-product pests.

NAMES

CHLORPHOXIM, BAYTHION-C SRA-7747

$$CH_3—CH_2—O(CH_3—CH_2—O)P(=S)—O—N=C(C \equiv N)—C_6H_4Cl$$

2-Chloro-alpha[(diethoxyphosphino-thioyloxy)-imino] phenylacetonitrile

TYPE: Baythion-C is an organic phosphate compound used as a contact insecticide.

ORIGIN: Bayer AG of Germany, 1980.

TOXICITY: LD_{50}-2500 mg/kg.

FORMULATIONS: 50% WP, 200 ULV.

PHYTOTOXICITY: Non-phytotoxic when used as directed.

USES: Outside the U.S. as a public health insecticide.

IMPORTANT PESTS CONTROLLED: Mosquitoes, both larvae and adults. Also controls simulium flies.

APPLICATION: Applied both indoors and outdoors where mosquitoes are located. May be applied by air.

PRECAUTIONS: Not for sale or use in the U.S. Toxic to some fish.

ADDITIONAL INFORMATION: Good residual activity. Low bird-toxicity.

NAMES

IODOFENPHOS, JODFENPHOS, NUVANOL-N

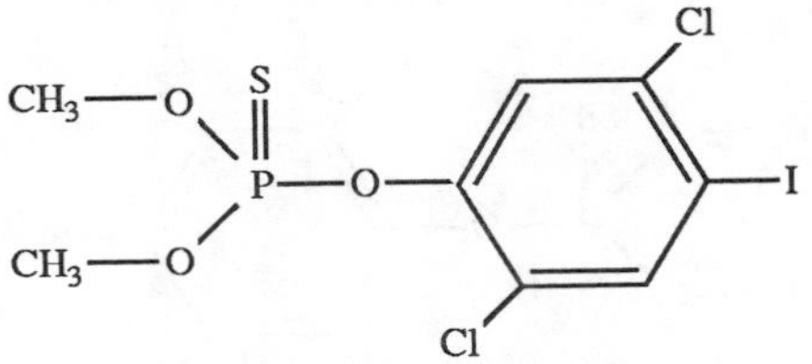

O-(2,5-Dichloro-4-iodophenyl) O,O-dimethyl phosphorothioate

TYPE: Iodofenphos is an organic phosphate compound used as an insecticide, killing both upon contact and as a stomach-poison.

ORIGIN: CIBA-Geigy Corporation, 1966.

TOXICITY: LD_{50}-2100 mg/kg.

FORMULATIONS: 50% WP, 2 lb/gal fogging concentrate, 2 EC, 5% dusts, 20% WP.

PHYTOTOXICITY: Non-phytotoxic when used as directed.

USES: Outside the U.S. as a public health insecticide, especially in Europe and the Middle East. Also used as a livestock insecticide.

IMPORTANT PESTS CONTROLLED: Lice, ticks, flies, mosquitoes, cockroaches, and others.

RATES: Applied at a .05-1% spray solution.

APPLICATION: Apply when insects appear and repeat as necessary.

PRECAUTIONS: Not for use in the U.S. Toxic to bees. Toxic to fish.

ADDITIONAL INFORMATION: Persists on treated surfaces for 3 months.

NAMES

***BROMOPHOS*, BROFENE, BROPHENE, KILSECT, NETAL, NEXION, OMEXAN**

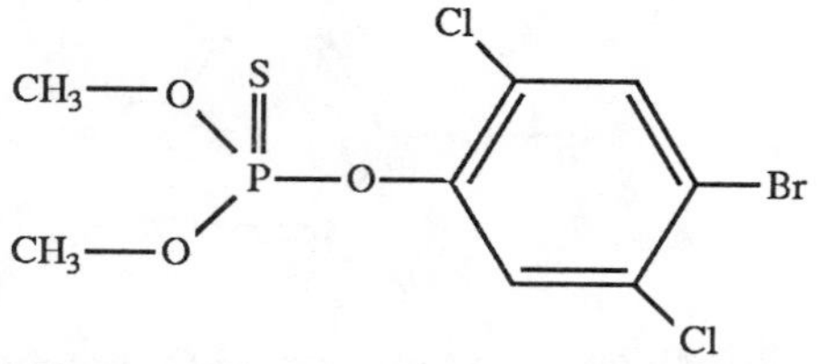

O,O-Dimethyl O,2,5-dichloro-4-bromophenyl-phosphorothioate

TYPE: Bromophos is an organic phosphate insecticide-acaricide, killing both as a contact and stomach-poison with residual effectiveness.

ORIGIN: C. H. Boehringer Sohn of Germany, 1964. Developed and marketed by Celamerck of Gmbh of West Germany.

TOXICITY: LD_{50}-3750 mg/kg.

FORMULATIONS: 2 EC and 4 EC, 25% and 40% WP, aerosols, 2% dust, 5% granules.

PHYTOTOXICITY: Do not use on cotton or grapes. Injury has been reported on varieties of cabbage, pears, and ornamentals. Repeated applications at short intervals should be avoided.

IMPORTANT PESTS CONTROLLED: Chinch bugs, flies, crickets, lice, mosquitoes, fruitflies, moths, fleas, root maggots, whiteflies, codling moths, beetles, aphids, thrips, cockroaches, caterpillars, and many others.

RATES: Usually used at a concentration of 250-1500 g active material/ha.

USES: Being used outside the U.S. on most field, vegetable, and fruit crops, as well as ornamentals, grain storage, and as a public health insecticide. Also used as a sheep-dip. Used also as a seed treatment.

APPLICATION: Spray infested areas thoroughly to ensure sufficient coverage for control. Repeat as necessary.

PRECAUTIONS: Do not use in the U.S. on any food or feed crop. Not compatible with sulfur and metal organic fungicides. Toxic to fish.

ADDITIONAL INFORMATION: A broad-spectrum insecticide with 1-6 weeks' residual effectiveness. Compatible with alkaline materials. Bees and many predators have some tolerance to this material. Appears to be synergistic with Lindane. Relatively harmless to wild life.

NAMES

BROMOPHOS-ETHYL, **FILARIOL, NEXAGAN, PARABAN**

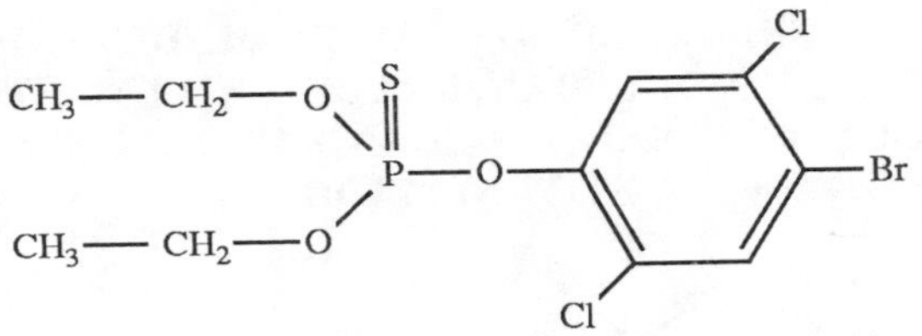

O,O-Diethyl O-2,5-dichloro-4-bromophenyl-phosphorothioate

TYPE: Bromophos-ethyl is an organic phosphate compound used as a selective insecticide-acaricide.

ORIGIN: C.H.Boehringer Sohn of Germany, 1964. Developed and marketed by Celamerck Gmbh of West Germany.

TOXICITY: LD_{50}-52 mg/kg.

FORMULATIONS: 40 and 80% EC, 5% granules.

IMPORTANT PESTS CONTROLLED: Mosquito larvae, as well as many agricultural pests. Also controls ticks and many other livestock pests.

USES: Agricultural premises, static-water tanks, septic tanks, cesspits, livestock, and as a crop-protection chemical on fruit, vegetables, field, and ornamental crops outside of the U.S. Also used as a mosquito larvicide.

RATES: Applied at 400-1200 g a.i./ha.

APPLICATION: Apply at 14-day intervals. Dip livestock as required.

PRECAUTIONS: Not registered for use in the U.S.

ADDITIONAL INFORMATION: Relatively Non-toxic to fish, birds, and reptiles. Not hazardous to livestock drinking water. A contact and stomach-poison with considerable residual effectiveness. Used as a dip in some countries on cattle, sheep, pigs, horses, camels, and dogs.

RELATED MIXTURES: Combat — A combination of bromophos-ethyl and diazinon developed by Celamerck to use as a fly spray outside the U.S.

NAMES

CHLORPYRIFOS, BRODAN, DETMOL, DETMOLIN, DOWCO 179, DURSBAN, ERADEX, KILLMASTER, LOCK-ON, LORSBAN, LOXIRAN, PYRINEX, SPANNIT, STIPEND, ZIDIL

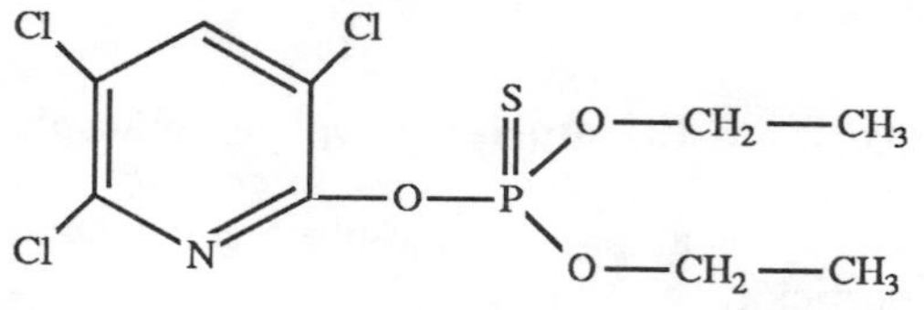

O,O-Diethyl-O-(3,5,6-trichloro-2-pyridyl) phosphorothioate

TYPE: Dursban is an organic phosphate insecticide which acts primarily by contact activity and as a stomach-poison.

ORIGIN: The Dow Chemical Company, 1966.

TOXICITY: LD_{50}-135 mg/kg.

FORMULATIONS: 4 EC, 2 EC, 15% granules, 30% flowable, 25% WP, and a 1/2% ready-to-use household spray.

PHYTOTOXICITY: Non-phytotoxic to most species of plants when used at the recommended rates. Poinsettias are injured by this material, as well as azaleas, camelias, roses, and variegated ivy.

IMPORTANT PESTS CONTROLLED: Ants, aphids, bagworms, boll weevil, bollworms, chinch bugs, cockroaches, corn borers, corn rootworm, crickets, cutworms, earwigs, grasshoppers, leafhoppers, mealybugs, mites, mosquitoes, peach tree borer, peach twig borer, scales, silverfish, spiders, thrips, ticks, tobacco budworms, webworms, whiteflies, white grubs, wireworms, and many others.

USES: Turf and premises, structural house pests, alfalfa, strawberries, beans, blueberries, collards, dates, filberts, kale, kolirabi, leeks, peas, pecans, cherries, mushrooms, asparagus, almonds, walnuts, grapes, figs, citrus, onions, soybeans, sunflowers, corn, peppers, sweet potatoes, sugar beets, tree fruits, tomatoes, cotton, ornamentals, and mosquito adults and larvae in the U.S. Used to control ticks on cattle and sheep in many foreign countries. Used to control numerous plant pests in other countries. Used as a seed treatment on corn and beans, soybeans, and some vegetables.

RATES: Applied at .1-5 lb actual/A.

APPLICATION: Use as a spot treatment to control household pests and as a wetting spray on lawns and ornamental plants. Applications for mosquito control are made with hand or power sprayers, mist applicators, or with aerial equipment. Mix with either oil or water for mosquito control. On turf, apply when temperature gets above 80°F and repeat at 2-month intervals. Applied to soil as a preplant or postemergence treatment. For foliar application, apply when insects appear and repeat as necessary. May be applied through spinkler irrigation systems.

PRECAUTIONS: Do not use in poultry houses, animal buildings, or where food is stored. Do not mix with alkaline compounds. Toxic to fish and crustaceans. Toxic to bees. Do not graze the treated areas.

ADDITIONAL INFORMATION: No systemic activity. Short reside on plant foliage. However, on soil, polluted water, wood, concrete, etc., it is effective for several weeks. Very resistant to leaching in the soil. Does not stain. Controls both the larvae and adult mosquito.

NAMES

CHLORPYRIFOS-METHYL, **DOWCO 214, GRAINCOTE, PYRINEX, RELDAN, SMITE, ZERTELL**

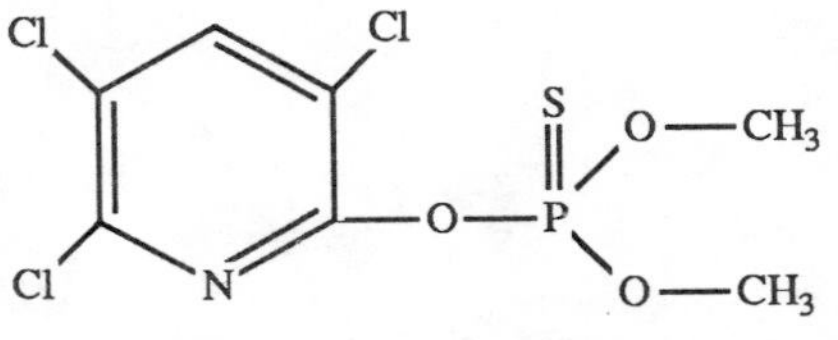

Cl Cl S O—CH_3 O—P Cl N O—CH_3

O,O-Dimethyl O-(3,5,6-trichloro-2-pyridyl) phosphorothioate

TYPE: DOWCO 214 is an organophosphate insecticide which kills by stomach, contact, and vapor action.

ORIGIN: The Dow Chemical Company, 1965. Being marketed in the U.S. by Gustafson.

TOXICITY: LD_{50}-1000 mg/kg.

FORMULATIONS: 2 EC, 4 EC, 25% WP, 1% G, 6 lb/gal oil-soluable concentrate.

PHYTOTOXICITY: **Non-phytotoxic** to most plant species when used at insecticidal rates.

IMPORTANT PESTS CONTROLLED: Various household and stored-products pests, flies, mosquitoes, and fruit and vegetable pests.

USES: Wheat, corn, sorghum, barley, rice and oats.Being used outside the U.S. for mosquito control, as a household insecticide, and on crop plants, such as vegetables, small grains, rice, sugar beets, cotton, citrus, grapes, and others and to control stored grain pests.

RATES: Applied at .005-.75 a.i./A. For storage apply at 180-340 ml (4E) per 1000 bushels.

APPLICATION: Use as a spot treatment to control household pests. Use conventional sprayers to obtain good coverage for control of plant pests. Use cold or thermal foggers, hand or power sprayers, mist applicators, or aerial equipment for control of mosquitoes. Apply by ground or air to cropland.

PRECAUTIONS: Not for sale or use in the U.S. Toxic to fish and crustaceans. Do not mix with alkaline compounds.

ADDITIONAL INFORMATION: No systemic activity. Short residue on plant foliage. Low dermal toxicity. Cholinesterase inhibitors. Does not persist in the soil. Broad-spectrum. Effective against both adults and larvae. Will control stored grain insects for over 1 year.

NAMES

COUMAPHOS, ASUNTOL, BAYMIX, CO-RAL, DIOLICE, MELDANE, MUSCATOX, RESITOX, UMBETHION

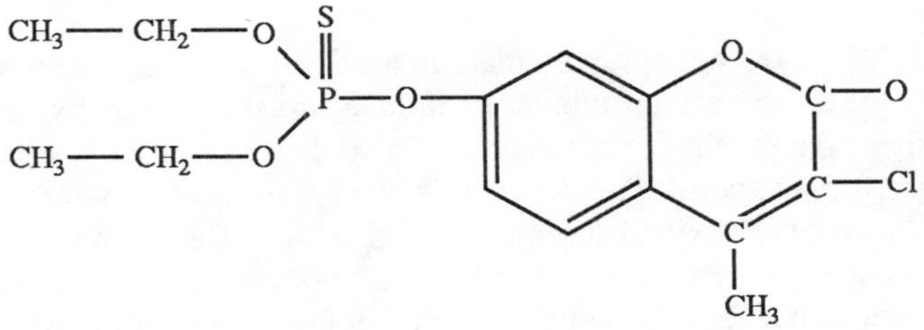

O,O-Diethyl-O-(3-chloro-4-methyl-2-oxo-2H-1-benzapyran-7-yl) phosphorothioate

TYPE: CO-RAL is a systemic, organic phosphate livestock insecticide.

ORIGIN: Bayer AG in Germany, 1956. Licensed to be sold in the U.S. by Cutter Labs, Div. of Miles Laboratories.

TOXICITY: LD_{50}-13 mg/kg.

FORMULATIONS: 25% WP, 1 and 5% dusts, 4% pour-on, 3% spray foam. 4.2 EC, 11.6% EC.

PHYTOTOXICITY: Generally not applied to plants.

USES: Registered for use on beef cattle, dairy cattle, sheep, dogs, goats, swine, horses, poultry, and agricultural premises.

IMPORTANT PESTS CONTROLLED: Grubs, flies, lice, ticks, keds, poultry mites, screwworms, mosquitoes, and others.

APPLICATION:

1. Backline treatment-Apply 1/2 oz/100 lb body weight for grub control.
2. Spray-Apply after heel fly season has passed at 250 psi or more pressure. Apply approximately 1 gal of the diluted spray per animal. Wet the skin, not just the hair.
3. Dip-Agitate the tank thoroughly prior to use. Repeat as necessary. Maintain adequate concentration in the vats.
4. Spot treatment-Use in infected wounds for screwworm control. May also be applied as a dust or in backrubbers.

5. Poultry house-Apply to litter beneath cages.
6. Backrubbers-Place backrubbers where animals travel daily. Refill as needed.
7. Cattle grubs-Treat at least 6 weeks before the expected appearance of the grubs in the back.

PRECAUTIONS: Do not spray animals in a confined, unventilated area. Do not apply to sick or stressed animals or animals less than 3 months old. Do not dip overheated animals. Do not treat within 10 days of shipping, weaning, vaccination, etc. Do not use before or after the application of natural or synthetic pyrethrins or compounds used to synergize them. Cattle on a fattening ration may be more subject to organic phosphate poisoning than animals on pasture or maintenance feed. Do not mix with other insecticides nor use in conjunction with oral drenches or other internal medicines.

ADDITIONAL INFORMATION: 10-20 day protection from screwworms can be obtained. Used to control fly larvae in poultry manure. Systemically controls cattle grubs in cattle. Used to control insects on humans.

NAMES

ETHOPROP, ETHOPROPHOS, MOCAP, PROPHOS

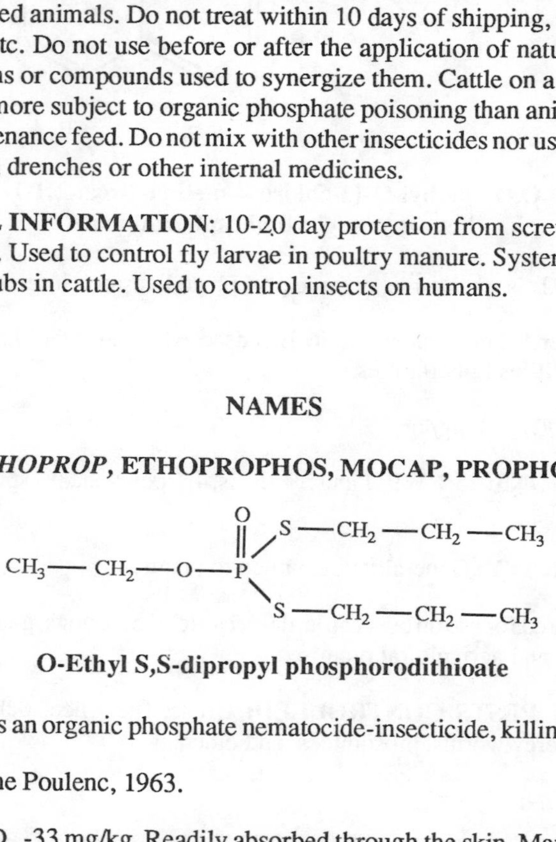

$$CH_3-CH_2-O-P(=O)(-S-CH_2-CH_2-CH_3)(-S-CH_2-CH_2-CH_3)$$

O-Ethyl S,S-dipropyl phosphorodithioate

TYPE: Mocap is an organic phosphate nematocide-insecticide, killing upon contact.

ORIGIN: Rhone Poulenc, 1963.

TOXICITY: LD_{50}-33 mg/kg. Readily absorbed through the skin. May cause eye and skin irritation.

FORMULATIONS: 10% granules, 6 EC.

PHYTOTOXICITY: Non-phytotoxic when used as directed.

USES: Corn, tobacco, mushrooms, peanuts, soybeans, pineapples, sugarcane, bananas, cabbage, cucumbers, beans, potatoes, sweet potaotes, turf, citrus seedlings, and ornamentals.

IMPORTANT PESTS CONTROLLED: Wireworms, nematodes, corn rootworms, sod webworms, Japanese beetles, cinch bugs, billbugs, and others.

RATES: Applied at 3/4-6 lb actual/A.

APPLICATION:

1. Corn-Apply in a 6-7 inch band over the row at planting time. Incorporate into the top 1/2 inch of soil immediately. Also may be applied by cultivating in, up to lay-by.
2. Tobacco-Apply and incorporate into the soil up to 2 weeks prior to planting. Incorporate into the top 4-6 inches of soil.
3. Nematode control-Usually applied and incorporated into the top 4-8 inches of soil prior to, or at, planting. May also be watered into the soil.

PRECAUTIONS: Incorporate into the soil as soon as possible. Toxic to wild life and fish. Do not use on newly seeded turf or on home lawns. Do not use as a seed furrow treatment.

ADDITIONAL INFORMATION: No fumigating activity. Kills upon contact only, although it does have residual activity of about 8 weeks. Most effective on soil-borne insects and nematodes.

RELATED MIXTURES:
Mocap-Plus: — A combination of Mocap and DiSystem produced by Mobil Chemical Co. for use on tobacco.

NAMES

OMETHOATE, BAY 45432, FOLIMAT, FOLIMATE

$$(CH_3-O)_2P(=O)-S-CH_2-C(=O)-NH-CH_3$$

O,O-Dimethyl S-(N-methylcarbamoylmethyl) phosphorothioate

TYPE: Folimat is an organic phosphate compound used as a systemic acaricide-insecticide.

ORIGIN: Bayer AG of Germany, 1965.

TOXICITY: LD_{50}-50 mg/kg. Irritating to the skin.

FORMULATIONS: 4 EC, 25 and 50% LC, 800 LC, and 1000 ULV.

PHYTOTOXICITY: May be phytotoxic to some peach varieties.

USES: Used outside the U.S. on deciduous fruits, grapes, citrus, forests, pastures, cereals, coffee, sugar beets, sugarcane, hops, cotton, potatoes, beets, vegetables, rice, ornamentals, and others.

IMPORTANT PESTS CONTROLLED: Mites, codling moth, gypsy moths, mealybugs, aphids, scale, whiteflies, thrips, leafhoppers, and many others.

RATES: Applied at .05 -.075% active ingredient concentration.

APPLICATION: Apply when insects first appear, and repeat as necessary.

PRECAUTIONS: Do not combine with sulfur. Do not use in the U.S. Harmful to bees.

ADDITIONAL INFORMATION: Especially effective on sucking insects. Systemic in activity. Compatible with most other pesticides. Apply during the same day it is mixed with water. The oxygen analog of dimethoate.

NAMES

VAMIDOTHION, KILVAL, TRUCIDOR

$$(CH_3-O)_2P(=O)-S-CH_2-CH_2-S-CH(CH_3)-C(=O)-NH-CH_3$$

O,O-Dimethyl S-(2-(1-methylcarbamoylethylthioethyl) phosphorothioate

TYPE: Vamidothion is an organic phosphate insecticide-acaricide, systemic in activity.

ORIGIN: Rhone-Poulenc of France, 1961.

TOXICITY: LD_{50}-64 mg/kg.

FORMULATION: 40% EC, 30% EC/ULV.

PHYTOTOXICITY: Non-phytotoxic at the recommended rates.

IMPORTANT PESTS CONTROLLED: Aphids, suckers, sawfly, thrips, leafhoppers, mites, whitefly, and others.

RATES: Applied at 6-8 oz actual/100 gal of water.

USES: Outside the U.S. on fruit crops, grapes, vegetables, field crops, beets, tobacco, hops, cotton, rice, and ornamentals.

APPLICATION: Apply when insects appear and repeat as necessary.

PRECAUTIONS: Do not use in the U.S., since there are no registrations.

ADDITIONAL INFORMATION: Systemic activity lasts for 3-8 weeks. No objectionable odor while this material is being used. Low toxicity to beneficial insects. Rapid acting. Translocation is chiefly upwards with the sap flow. Outstanding activity against the wooly aphid.

NAMES

***TRICHLORONAT*, AGRISIL, AGRITOX, BAY 37289, FENOPHOSPHON, PHYTOSOL, TRICHLORONAT**

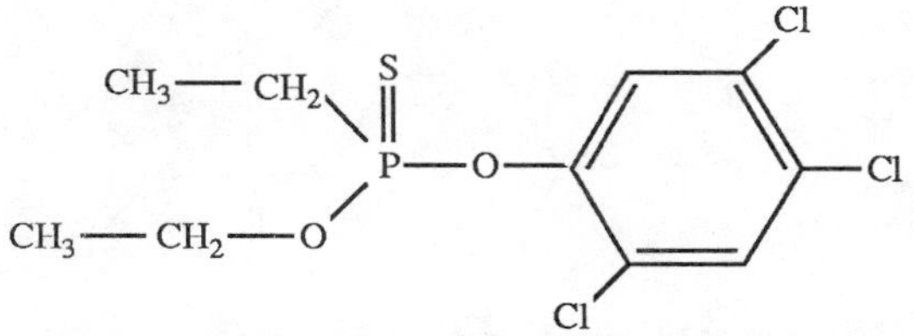

O-Ethyl O-(2,4,5-trichlorophenyl) ethylphosphonothioate

TYPE: Agritox is an organic phosphate compound, especially effective on soil insects, with good contact and stomach-poison activity.

ORIGIN: Bayer AG of Germany, 1960.

TOXICITY: LD_{50}-16 mg/kg.

FORMULATIONS: 50 % EC, 2.5 and 7% granules, 20% seed-dressing powder.

PHYTOTOXICITY: Non-phytotoxic when used as directed.

USES: Outside the U.S. on turf, cole crops, corn, cereals, onions, grasslands, carrots, and bananas. Controls soil pests in meadows.

IMPORTANT PESTS CONTROLLED: Wireworms, root maggots, onion fly, carrot fly, aphids, lygus, leafhoppers, grasshoppers, symphilids, corn earworms, sod webworm, and many others.

RATES: Applied at 1/2-3 lb actual/A.

APPLICATION: Apply as a soil insecticide, either by soil incorporation or as a water drench. Also may be used as a seed treatment.

PRECAUTIONS: Not to be used in the U.S. Do not mix with alkaline compounds. Do not use to treat seed previously treated with mercurial compounds.

ADDITIONAL INFORMATION: Considered most effective on soilborne insects. Soil applications have given up to 5 months' residual control. Compatible with most insecticides and fungicides.

NAMES

***FAMPHUR*, BO-ANA, CYFLEE, DAVIP, FAMOPHOS, FANFOS, WARBEX**

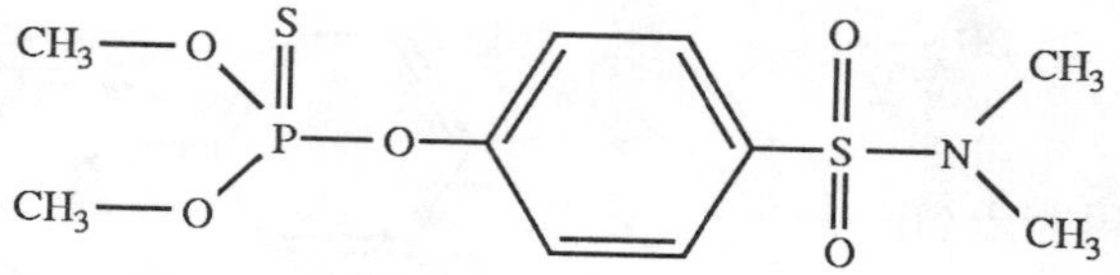

O, O-Dimethyl O-p-(dimethylsulfamoyl) phenyl phosphorothioate

TYPE: Famphur is an organic phosphate compound being used as a systemic livestock insecticide.

ORIGIN: American Cyanamid Company, 1966.

TOXICITY: LD_{50}-35 mg/kg.

FORMULATIONS: Liquid 13.2%, $33^{1}/3$% feed premix.

PHYTOTOXICITY: Not normally used on plants.

USES: Cattle.

IMPORTANT PESTS CONTROLLED: Heel flies and grubs. The migrating larvae are effectively killed. Cattle lice.

RATES: Apply at the rate of 1 oz/200 lb body weight with a maximum dosage of 4 oz. As a feed premix, apply at 2-1/2-10 mg/kg body weight for 5-30 consecutive days, depending upon the pest to be controlled.

APPLICATION: Used as either a liquid pour-on application or mixed with feed for the animals. Apply as soon as possible after heel fly activity ceases. This means not after October 1 in certain areas of the Southwest, and not after November 1 in the rest of the U.S. Do not apply for a few days if animals will get wet due to adverse weather conditions.

PRECAUTIONS: Do not use on sick or weak livestock. Do not slaughter livestock within 35 days after treatment. Do not use on calves less than 3 months old. Do not use on lactating dairy cows or dairy cattle within 21 days of freshening. Brahman are less tolerant to this material than other breeds. Eliminate swine from area where run off occurs.

ADDITIONAL INFORMATION: Most effective against the larvae of the Hypoderma spp. of flies. Systemic in activity. Controls the migrating larvae where administered either orally or dermally. Treatment of grubs within 4-6 weeks of emergence is of little value.

NAMES

CYANOPHOS, CYANOX, CYAP, CYNOCK, S-4084

O,O-Dimethyl O-(4-cyanophenyl) phosphorothioate

TYPE: Cyanox is an organophosphate compound used as a foliar insecticide.

ORIGIN: Sumitomo Chemical Co. of Japan, 1960.

TOXICITY: LD_{50}-610 mg/kg.

FORMULATIONS: 50% EC, 3% dust, 1% oil-based liquid spray, 5% EC.

PHYTOTOXICITY: Considered Non-phytotoxic when used as directed.

USES: Used in Japan on cabbage, fruit trees, radish, sweet potatoes, sugar beets, egg plant, potatoes, soybeans, vegetables, and for sanitary uses.

IMPORTANT PESTS CONTROLLED: Armyworms, diamondback moth, aphids, fleabeetles, houseflies, mosquitoes, cockroaches, bedbugs, and others.

RATES: Applied at .025-.1% concentration.

APPLICATION: Apply when insects appear, and repeat as necessary.

PRECAUTIONS: Mites are not controlled. Moderately toxic to fish. Toxic to honeybees. Not used in the U.S.

ADDITIONAL INFORMATION: Quick knockdown effects are observed with this material. First used commercially in 1967.

NAME

FENPHOSPHORIN, SALITHION

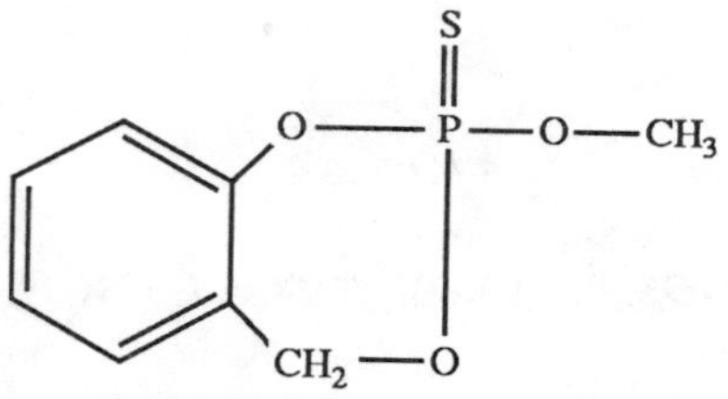

2-Methoxy-4H-1,3,2-benzodioxaphosphorine-2-sulfide

TYPE: Salithion is an organic phosphate compound used as a foliar-contact insecticide.

ORIGIN: Pesticide Research Lab. of the University of Kyushu in Japan, 1962. Licensed to be marketed and developed by Sumitomo Chemical Co. of Japan.

TOXICITY: LD_{50}-125 mg/kg.

FORMULATIONS: 25% EC, 25% WP, 5 and 10% granules.

PHYTOTOXICITY: Non-phytotoxic when used as directed.

USES: Used in Japan on apples, cotton, tea, tobacco, grapes, rice, cotton, pears, vegetables, and others.

IMPORTANT PESTS CONTROLLED: Riceborers, leafhoppers, planthoppers, cutworms, fruit flies, leaf rollers, scales, bollworms, boll weevils, aphids, Oriental fruitmoth, armyworms, cabbageworms, diamondback moth, and others.

RATES: Applied at .170-500 ppm a.i.

APPLICATION: Apply when insects first appear, and repeat as necessary.

PRECAUTIONS: Not for sale or use in U.S. Toxic to fish.

ADDITIONAL INFORMATION: First used commercially in 1968. Do mix with strong alkaline materials.

NAME

PYRACLOFOS, TIA-230, BOLTAGE

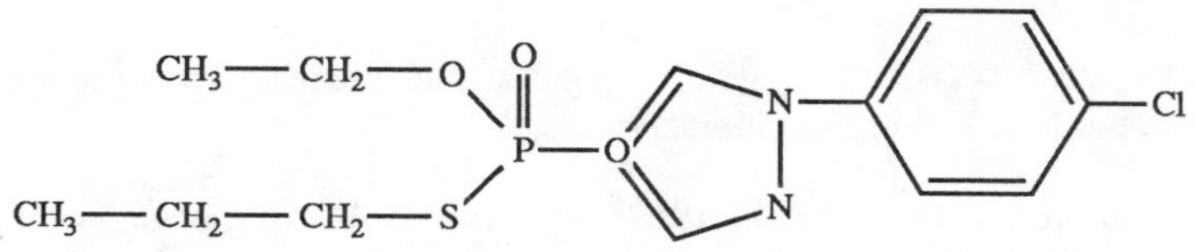

CH_3— CH_2—O, O, P—O, N, N, Cl

CH_3— CH_2— CH_2—S

O-Ethyl S-N-propyl O-(1-p-chlorophenyl)O-pyrazoyl phosphorothioate

TYPE: TIA-230 is an organic phosphate compound used as a contact and stomach-poison insecticide-acaricide.

ORIGIN: Takeda Chemical Ind. of Japan, 1982.

TOXICITY: LD_{50}-237 mg/kg.

FORMULATIONS: 50% EC, 35% WP.

PHYTOTOXICITY: Some injury has been noted on apples, pears, peaches, and citrus.

IMPORTANT PESTS CONTROLLED: Thrips, diamondback moth, oriental fruit moth, corn borer, budworms, bollworms, armyworms, loopers, cutworms, rice weevil, whitefly, aphids, mites, and others.

USES: Tea, vegetables, peas, rice, soybeans, cotton, tobacco, sugarbeets, potatoes, corn, and others outside the U.S.

RATES: Applied at .5-1 kg. a.i./ha.

APPLICATION: Apply when insects appear and repeat as necessary.

PRECAUTIONS: Not for use in the U.S. Toxic to fish.

NAMES

***METHAMIDOPHOS*, BAY 71628, HAMIDOP, MONITOR, TAMARON**

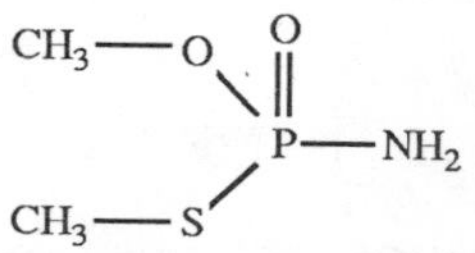

$$CH_3{-}O{-}P({=}O)(NH_2){-}S{-}CH_3$$

0, S-Dimethyl phosphoramidothioate

TYPE: Methamidophos is a systemic, organic phosphate compound used as a systemic, residual insecticide-acaricide.

ORIGIN: Chevron Chemical Co., 1967. Also Bayer AG of Germany who have licensed Mobay Chemical Corp. to develop and market in the U.S. Sold in the U.S. by Valent Chemical Co. also.

FORMULATIONS: 4 EC, 6 EC, 25% WP, 5% granules, 600 g/1 EC.

TOXICITY: LD_{50}-29.9 mg/kg.

PHYTOTOXICITY: Non-phytotoxic when used as directed.

USES: Alfalfa, broccoli, Brussels sprouts, cabbage, cauliflower, celery, clover, sugar beets, cotton, and potatoes. Used on a number of these in other countries, as well as vegetables, hops, corn, peaches, and other crops.

IMPORTANT PESTS CONTROLLED: Aphids, flea beetles, worms, whiteflies, cabbage looper, thrips, cutworms, Colorado potato beetle, potato tuberworms, armyworms, mites, leafhoppers, and many others.

APPLICATION: Apply when insects appear, and repeat as necessary, usually at 7-10 day intervals.

PRECAUTIONS: Slightly corrosive to mild steel and copper alloys. Do not mix with alkaline compounds. Toxic to birds and other wildlife. Do not graze treated areas. Toxic to bees.

ADDITIONAL INFORMATION: Besides being a contact and stomach-poison, it processes systemic activity. Compatible with most other pesticides. Good residual effectiveness.

NAMES

ACEPHATE, KITRON, ORTHENE, ORTRAN, ORTRIL, TORNADO

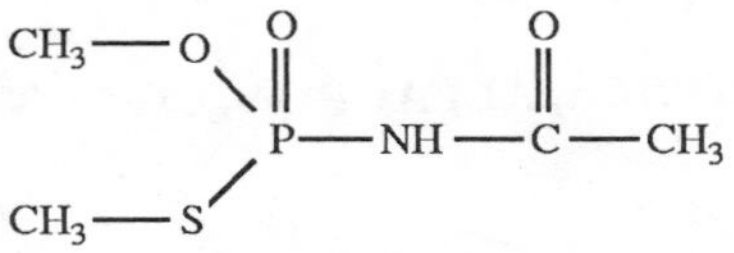

```
CH3—O   O            O
     \  ‖            ‖
      P — NH — C — CH3
     /
CH3—S
```

O, S-Dimethyl acetylphosphoramidothioate

TYPE: Orthene is an organic phosphate compound used as a contact and systemic insecticide.

ORIGIN: Chevron Chemical Co. 1967. Being developed in Japan by Hokko Chemical Co. and Takeda Chemical Co. Being sold in the U.S. by Valent Chemical Co.

TOXICITY: LD_{50}-866 mg/kg.

FORMULATIONS: 75% SOLUABLE POWDER, 50% WP, 2% DUST.

PHYTOTOXICITY: A marginal leaf-burn has occurred on Red Delicious apples, otherwise considered Non-phytotoxic. Do not apply to American elm, Flowering crab apple, sugar maple, cottonwood, or huckleberry.

USES: Ornamental trees, beans, cauliflower, Brussel sprouts, celery, cotton, cranberries, grasses, lettuce, macadamia nuts, peanuts, peppers, rangeland, pastures, soybeans, tobacco, shrubs, and flowers.

IMPORTANT PESTS CONTROLLED: Aphids, armyworms, bagworms, grasshoppers, cabbage looper, budworms, diamondback moth, leaf miners, gypsy moths, caterpillars, cabbageworm, bollworm, lygus bugs, leafhoppers, cutworms, thrips, and many others.

RATES: Applied at 1/2-1 lb actual/A.

APPLICATION: Apply when insects appear, and repeat as necessary. Used as a seed treatment on cotton.

PRECAUTIONS: Do not graze treated areas. Toxic to bees. Avoid drift.

ADDITIONAL INFORMATION: An insecticide of moderate persistence, with residual activity of 6-9 days. Relatively Non-toxic to fish. May be combined with other pesticides.

NAMES

AZAMETHIPHOS, ALFACRON, CGA-18809, SNIP

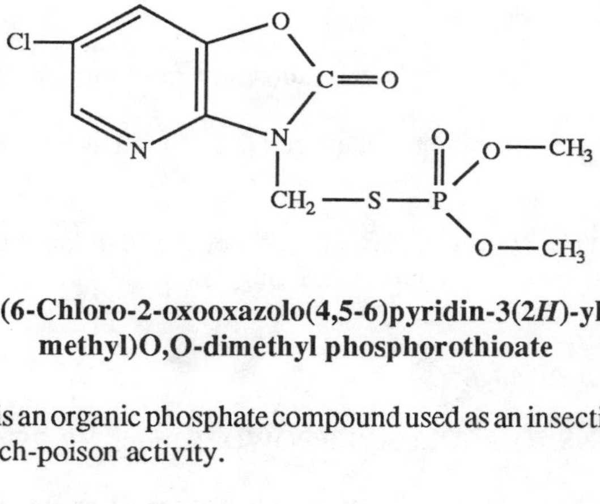

s-(6-Chloro-2-oxooxazolo(4,5-6)pyridin-3(2*H*)-yl) methyl)O,O-dimethyl phosphorothioate

TYPE: Alfacron is an organic phosphate compound used as an insecticide with contact and stomach-poison activity.

ORIGIN: CIBA-Geigy, 1978.

TOXICITY: LD_{50}-1180 mg/kg. May cause slight eye irritation.

FORMULATIONS: 50% WP, 10% WP, 2 EC.

PHYTOTOXICITY: None noted to date.

USES: Being used outside the U.S. for public health control and farm fly control.

IMPORTANT PESTS CONTROLLED: Mosquitoes, flies, etc.

RATES: Applied at .25-.37 lb a.i./100 gal of water or at .5-1 lb a.i./A.

APPLICATION: Apply when insects appear, and repeat as necessary.

PRECAUTIONS: Not for sale or use in the U.S. Toxic to fish.

ADDITIONAL INFORMATION: A broad-spectrum insecticide. May be tank mixed with other pesticides.

NAMES

METHACRIFOS, **CGA-20168, DAMFIN**

CH_3—O, S, P—O, CH_3, CH_3—O, C=C, (E)—isomer, H, C=O, CH_3

O-2-methoxy carbonyl-1-emyl O,O-dimethyl phosphorothioate

TYPE: Methacrifos is an organic phosphate compound used as a contact and vapor-action insecticide.

ORIGIN: CIBA-Geigy, 1978.

TOXICITY: LD_{50}-678 mg/kg.

PHYTOTOXICITY: Not to be used on plant foliage.

USES: Being used to control stored-products insects.

RATES: Applied at 5-15 ppm to stored grains.

IMPORTANT PESTS CONTROLLED: Weevils, beetles, moths, and many others.

APPLICATION: Apply where insects frequent. Also applied to grain as it goes into storage by mixing or as a surface treatment.

PRECAUTIONS: Not for sale or use in the U.S.

ADDITIONAL INFORMATION: Has a quick knockdown effect. provides several months' protection to stored grains.

NAMES

SD-208304, FORTRESS

$$CH_3-CH_2-O-\overset{\overset{S}{\|}}{P}-O-\overset{\overset{Cl}{|}}{CH}-\underset{\underset{Cl}{|}}{\overset{\overset{Cl}{|}}{C}}-Cl$$

***O.,O*-diethyl *O*, 1,2,2,2-tetrachloroethyl phosphorothioate**

TYPE: Fortress is an organic phosphate compound used as a broad spectrum soil insecticide.

ORIGIN: DuPont Co., 1984.

TOXICITY: LD_{50} 1-10 mg/kg. May cause slight eye and skin irritation.

FORMULATION: 15% granule.

PHYTOTOXICITY: Non-phytotoxic when used as directed.

USES: Experimentally being used on corn, potatoes, sugarbeets, sorghum, vegetables, turf and other crops.

IMPORTANT INSECTS CONTROLLED: Corn rootworm, cutworms, European corn borer, wire worms, root maggot and others.

RATES: Applied at .56-1.1 kg. a.i./ha.

APPLICATION: Apply the granule in the seed furrow at time of planting. May also be applied broadcast and incorporated into the soil.

PRECAUTION: To be used on an experimental basis only.

ADDITIONAL INFORMATION: Low water solubility so it does not move very far in the soil. Broad spectrum. Control lasts throughout the season.

ORGANIC PHOSPHATES
Dithiophosphate Prototypes

NAMES

***MALATHION,* CARBOPHOS, CHEMATHION, CYTHION, EMMATOS, FYFANON, KARBOFOS, KYPFOS, MALAMAR, MALAPHOS, MALASPRAY, MALPHOS, MERCAPTOTHION, MLT, ZITHIOL**

```
               S                              O
CH3—O          ‖                              ‖
      \        ‖                              ‖
        P—S—CH—CH2—C—O—CH2—CH3
      /            |
CH3—O              C—O—CH2—CH3
                   ‖
                   O
```

O,O-Dimethyl phosphorodithioate ester of diethyl mercaptosuccinate

TYPE: Malathion is an organic phosphate insecticide-acaricide.

ORIGIN: American Cyanamid, 1950.

TOXICITY: LD_{50}-1375 mg/kg.

FORMULATIONS: 25 and 50% WP, 4, 5, 7, and 9.7 EC, 4 and 5% dusts, 1, 2, and 4% aerosols, granules, 5 and 10% baits.

PHYTOTOXICITY: Injury has been reported on McIntosh and Cortland varieties, of apples, as well as sweet cherries, certain European grapes, Bosc pears, cucurbits, and string beans. Fruit spotting has resulted on nectarines. Certain ornamentals have been injured, such as ferns, hickory, viburnum, lantana, Crassula and Canaerti junipers, petunias, spirea, white pines, maples, and elms.

USES: Alfalfa, almonds, anise, apples, apricots, asparagus, avocados, barley, beans, beets, blackeye peas, blueberries, caneberries, broccoli, Brussels sprouts, cabbage, carrots, cauliflower, citrus, cowpeas, celery, cherries, chestnuts, clover, collards, corn, cotton, cucumbers, currants, dandelion, dates, dewberries, eggplant, endive, figs, filberts, flax, garlic, gooseberries, grapes, grasses, guavas, hops, horseradish, kale, kohlrabi, kumquats, leeks, lentils, lespedeza, lettuce, macadamia nuts, mangoes, melons, mint, mushrooms, mustard, nectarines, oats, okra, onions, parsley, papayas, parsnips, passion fruit, peaches, peanuts, pears, peas, pecans, peppers, pineapples, potatoes, plums, prunes, pumpkins, quince, radishes, rice, rutabagas, rye, safflower, shallots, sorghum, soybeans, spinach, squash, strawberries, sugar beets, Swiss chard, sweet potatoes, tobacco, tomatoes, turnips, vetch, walnuts, watercress, wheat, wild rice, cattle, poultry, sheep, goats, swine, greenhouses, general fogging, agricultural premises, poultry ranges, stored grains, grain bins, homes, ornamentals, forest insect control, and mosquito control.

IMPORTANT PESTS CONTROLLED: Aphids, mites, scale, flies, leafhoppers, leaf miners, thrips, loopers, pear psylla, mealybugs, Japanese beetles, lygus bugs, spittlebugs, corn earworms, chinch bugs, grasshoppers, armyworms, boll weevils, bollworms, lice, ticks, ants, spiders, mosquitoes, and many others.

RATES: Applied at 1/2-2 lb actual/100 gal of water or 1/2 -3 lb actual/A.

APPLICATION:

1. Foliage-Apply at a uniform rate with common application equipment. Repeat as necessary.
2. Soil-Wash into the top 6-8 inches of soil, or leave on the soil surface around the plants.
3. Livestock-Dip, spray, or dust individual animals or birds thoroughly. Do not apply to dairy animals within 5 hours of milking time.
4. Seed and grain storage-Treat seed with common seed treaters. Apply to grain as it goes into storage.
5. Premises-Applied as a spray or brush on side of buildings. May also be used as a wet or dry bait. May be applied to ant mounds and to uncultivated land for grasshopper control.

PRECAUTIONS: Reacts with heavy metals, especially iron. Incompatible with alkaline materials. The ULV concentrate may cause spotting on automobile paint finish. Wash immediately if automobiles are accidently sprayed. Toxic to fish. Toxic to bees.

ADDITIONAL INFORMATION: Compatible with most insecticides and fungicides. When mixed with alkaline materials, initial kills are satisfactory, but residual toxicity may be decreased. Does not persist in the soil.

NAMES

***AZINPHOS-METHYL*, AZITOX, CARFENE, COTNION-METHYL, CRYSTHYON, GOTHNION, GUSATHION, GUTHION, METILTRIAZOTION**

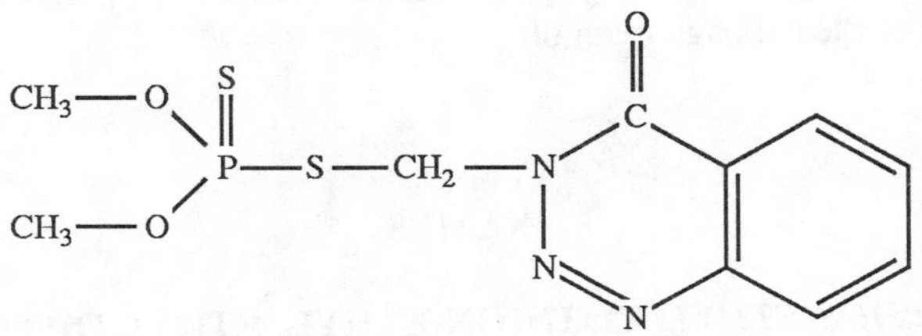

O,O-Dimethyl S-(4-oxo-1,2,3-benzotriazin-3 (4H)-ylmethyl) phosphorodithioate

TYPE: Guthion is an organic phosphate insecticide-acaricide used as a stomach and contact poison.

ORIGIN: Bayer AG in Germany, 1954. Sold in the U.S. by Mobay Chemical Corp.

TOXICITY: LD_{50}-4.4 mg/kg. Absorbed through the skin.

FORMULATIONS: 2EC, 25% and 35% WP, 5%dust.

PHYTOTOXICITY: Injury to Hawthorn and American Linden has occurred. The EC has caused fruit russeting to some fruit varieties.

USES: Alfalfa, almonds, apricots, apples, artichokes, beans, barley, caneberries, blueberries, broccoli, Brussels sprouts, cabbage, cauliflower, celery, cherries, citrus, clover, cotton, crab apples, cranberries, cucumbers, eggplant, filberts, grapes, gooseberries, kiwi, melons, nectarines, oats, onions, pastures, peaches, pears, pecans, plums, peppers, potatoes, prunes, quinces, rye, soybeans, spinach, strawberries, sugarcane, tobacco, tomatoes, walnuts, watermelons, wheat, and ornamentals. Widely used outside of the U.S. on these and coffee, forestry, pastures, mangoes, rice, sugar beets, and other crops.

IMPORTANT PESTS CONTROLLED: Aphids, mites, codling moths, lygus bugs, bollworms, armyworms, boll weevils, thrips, grasshoppers, stinkbugs, spittlebugs, plum curculio, and many others.

RATES: Applied at 1 oz-1 lb actual/100 gal of water or 4-6 lb actual/A.

APPLICATION: Apply at a uniform rate in power-operated ground and aircraft sprayers. Repeat as necessary.

PRECAUTIONS: Do not feed treated crop residue to livestock. Incompatible with Bordeaux, lime sulfur, and lime. Highly toxic to fish and wildlife. Do not use on greenhouse crops. Harmful to bees.

ADDITIONAL INFORMATION: Residual effects of 1-3 weeks. Compatible with other pesticides. Sometimes applied to the furrow before planting cole crops for control of the cabbage maggot.

NAMES

AZINPHOS-ETHYL, COTNION-ETHYL, ETHYL GUTHION, GUSATHION A, TRIAOTION

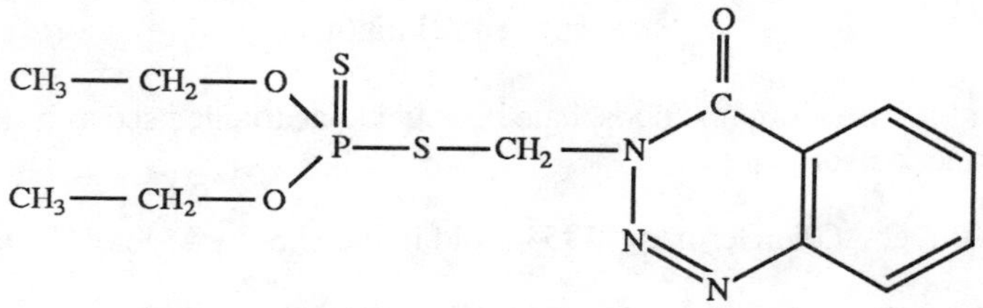

O,O-Dimethyl-S-((4-oxo-3H-1,2,3-benzotriazine-3-yl)-methyl)-dithiophosphate

TYPE: Ethyl Guthion is an organic phosphate insecticide-acaricide with a long residual activity, acting as both a contact and stomach-poison.

ORIGIN: Bayer AG of Germany, 1953.

TOXICITY: LD_{50}-12 mg/kg.

FORMULATIONS: 25 and 40% EC, 25-40% WP, 500 g/1 ULV.

PHYTOTOXICITY: Considered to be Non-phytotoxic when used as recommended. Use WP formulations on fruit trees to prevent russetting.

USES: In other countries it is widely used, especially on fruit and vegetable crops, cotton, forestry, pastures, coffee, cereals, potatoes, hops, grapes, citrus, rice, tobacco, and others.

IMPORTANT PESTS CONTROLLED: Spider mites, aphids, caterpillars, potato bug, beetles, boll weevils, whiteflies, bollworms, thrips, and most other sucking or biting insects.

RATES: Used at 1/4-3/4 lb actual/100 gal of water or at 1/4-1/2 lb actual/A.

APPLICATION: Apply before pests build up into destructive numbers. Apply evenly and thoroughly, repeating as necessary.

PRECAUTIONS: Not used in the U.S. Toxic to bees and fish.

ADDITIONAL INFORMATION: Nonsystemic. More effective against red spider mite than Guthion. However, Guthion has proven somewhat superior on certain other insects. Controls mites in all stages, even with some ovicidal activity. Long residual activity. Broad-spectrum activity. Compatible with most other pesticides.

NAMES

MEPHOSFOLAN, CYTROLANE

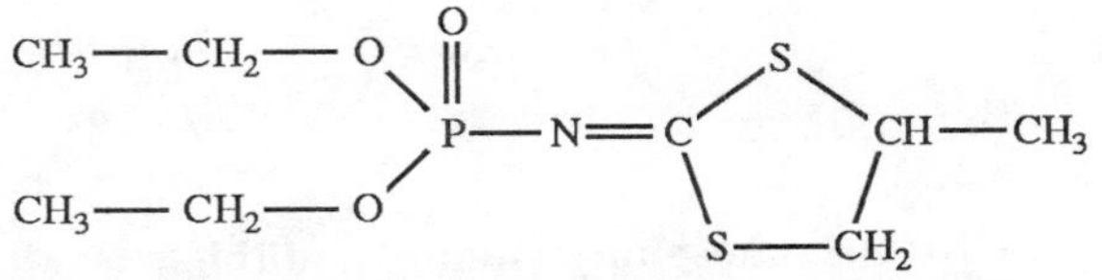

$CH_3—CH_2—O$ O S
P—N=C CH—CH_3
$CH_3—CH_2—O$ S—CH_2

2-(Diethoxyphosphinylimino)-4-methyl-1,3-dithiolane

TYPE: Cytrolane is an organic phosphate insecticide-acaricide with contact, stomach-poison, and systemic activity.

ORIGIN: American Cyanamid Company, 1962.

TOXICITY: LD_{50}-3 mg/kg.

FORMULATIONS: 2% and 10% granules, 25 and 50% EC.

PHYTOTOXICITY: Cotton foliage has been injured when grown under high-rainfall conditions.

USES: Used outside the U.S. on cotton, citrus, tobacco, corn, sorghum, rice, potatoes, sugar beets, tomatoes, sugarcane, and other crops.

IMPORTANT PESTS CONTROLLED: Corn borers, alfalfa weevils, cotton leafworms, fleahoppers, boll weevils, bollworms, mites, loopers, leaf miners, cutworms, armyworms, rootworms, hornworms, aphids, grasshoppers, lygus bugs, leafhoppers, and many others.

RATES: Applied at .1-1 lb a.i./A.

APPLICATION: Applied to both the soil and the foliage. Side dressing or granules over the seed or plant rows are methods of soil treatments. Treat foliage before pests build up to threatening numbers. Repeat as necessary.

PRECAUTIONS: Not for use in the U.S.

ADDITIONAL INFORMATION: Relatively long residual activity. May be used both as a foliar and soil insecticide.

NAMES

PHOSFOLAN, CYLAN, CYOLAN, CYOLANE

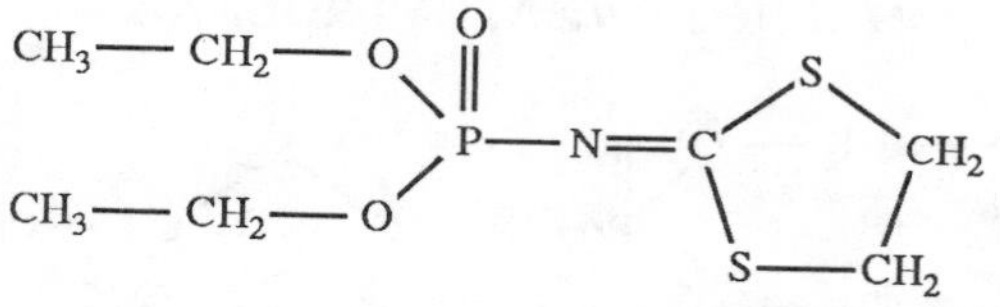

O
CH_3— CH_2— O
S
P— N=C
CH_2
CH_3— CH_2— O
S— CH_2

2-(Diethoxyphosphinylimino)-1,3-dithiolane

TYPE: Cyolane is an organic phosphate insecticide-acaricide with contact, stomach-poison, and systemic activity.

ORIGIN: American Cyanamid Company, 1962.

TOXICITY: LD_{50}-8.9 mg/kg.

FORMULATIONS: 25% EC, 10% granules.

PHYTOTOXICITY: Non-phytotoxic when used at the recommended rates.

USES: Used outside the U.S. on cotton, tobacco, onions, cabbage, and other crops.

IMPORTANT PESTS CONTROLLED: Leafhoppers, fleahoppers, cotton leafworm, boll weevils, mites, loopers, leaf miners, cutworms, armyworms, rootworms, hornworms, aphids, grasshoppers, lygus bugs, and many others.

RATES: Applied at .1-1 lb actual/A.

APPLICATION: Applied both to the soil and to the foliage. Seed dressing or granules over the seed or plant rows are methods of soil treatments. Treat foliage before pests build up to threatening numbers. Repeat as necessary.

PRECAUTIONS: Not for use in the U.S.

ADDITIONAL INFORMATION: Relatively long residual activity, giving control for up to 21 days. Effective both as a foliar insecticide and as a soil insecticide. Quickly taken up by the plant.

NAMES

DISULFOTON, THIODEMETON, DIMAZ, DI-SYSTON, DITHIODEMETON, DITHIOSYSTOX, FRUMIN-AL, KNAVE, SOLVIREX

$$(CH_3-CH_2-O)_2P(=S)-S-CH_2-CH_2-S-CH_2-CH_3$$

O,O-Diethyl-S-2-ethylthioethyl-phosphorodithioate

TYPE: Di-Syston is a selective, systemic, organic phosphate insecticide-acaricide.

ORIGIN: Bayer AG in Germany, 1956. Licensed to be sold in the U.S. by Mobay Chemical Corporation. Also sold by Sandoz Ltd. outside the U.S.

TOXICITY: LD_{50}-2.6-12.5 mg/kg. Absorbed through the skin.

FORMULATIONS: 6 EC, 5-15% granules, 50% WP, seed-dressing powder.

PHYTOTOXICITY: High dosages may injure seed. Some leaf burn to alfalfa has occurred. Garden lily bulbs have been injured.

USES: Alfalfa, asparagus, barley, beans, broccoli, Brussels sprouts, cabbage, cauliflower, clover, coffee, cotton, corn, hops, lettuce, oats, peanuts, pecans, pineapple, potatoes, rice, strawberries, sugar beets, soybeans, spinach, sugarcane, tobacco, sorghum, tomatoes, wheat, ornamentals, non-bearing fruit trees, and forest plantings. Used outside the U.S. on a number of crops.

IMPORTANT PESTS CONTROLLED: Aphids, mites, thrips, leafhoppers, flea-beetles, lace bugs, leaf rollers, whiteflies, mealybugs, leaf miners, Mexican bean beetles, and many others.

RATES: Applied at 1/2-3 lb actual/A.

APPLICATION: Applied as a soil insecticide by drilling, side dressing or broadcasting. It can be applied preplant, preemergence, or postemergence. On fruit trees, spread from the trunk to the drip line beneath the tree. Work into the soil and water thoroughly. Also applied as a seed coating and as a mixture with dry or liquid fertilizer. Used on some crops as a foliar insecticide. Used as a seed treatment on cotton.

PRECAUTIONS: Do not apply within direct contact of the seed, except the seed dressing formulation. On cotton, sugar beets, tomatoes, beans, or lettuce, the maximum recommended rate may cause some delay of emergence and stunting under conditions of cool, wet weather. Injury may be more pronounced on light, sandy soils. Keep all unprotected persons out of the area until the residues have dissipated. Toxic to fish. Do not use in greenhouses. Toxic to bees. Plant injury may be enhanced when used with certain preemergence herbicides.

ADDITIONAL INFORMATION: Compatible with most pesticides. Absorbed through the roots. Control of up to 7 weeks may be obtained.

NAMES

THIOMETON, DITHIOMETHON, EBICID, EKATIN, LUXISTELM, TOMBEL

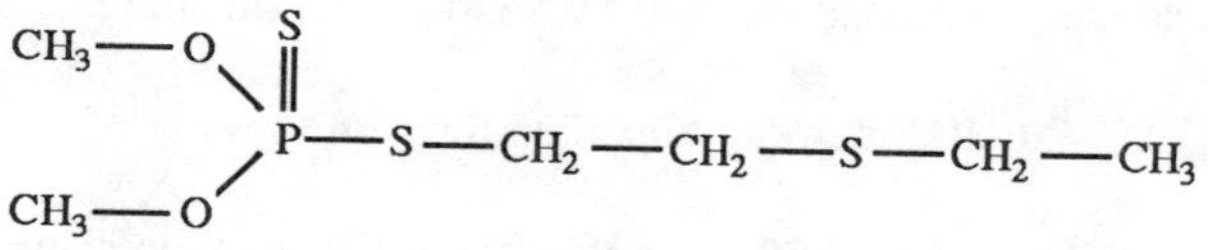

$$\begin{array}{l} CH_3{-}O \quad \overset{S}{\|} \\ \qquad\qquad P{-}S{-}CH_2{-}CH_2{-}S{-}CH_2{-}CH_3 \\ CH_3{-}O \end{array}$$

S-2-Ethylthioethyl O,O-dimethyl phosphorodithioate

TYPE: Thiometon is an organic phosphate insecticide-acaricide that possesses systemic activity.

ORIGIN: Sandoz Ltd. of Switzerland, 1953.

TOXICITY: LD_{50}-85 mg/kg.

FORMULATIONS: 25% EC, 15% ULV.

PHYTOTOXICITY: Considered Non-phytotoxic, except on such ornamentals as ferns, cyclamen, Baccarat, Roslandia, and Polyantha roses under glass.

USES: Used outside the U.S. as a foliar spray on many crops and ornamentals.

IMPORTANT PESTS CONTROLLED: Aphids, mites, psyllids, sawflies, thrips, and many other sucking insects.

RATES: Apply at a .1% concentration.

APPLICATION: For the maximum systemic activity, application should be made when the plant is actively growing. Treat as early as possible to prevent an insect population buildup. The systemic action will kill the hidden insects in 2-3 days. Repeat when necessary. May also be used as a soil drench.

PRECAUTIONS: Do not mix with strong alkaline compounds. Not sold in the U.S.

ADDITIONAL INFORMATION: Control may last for 10-20 days. May be applied in the irrigation water where it is taken up in the plants' root system. Translocated within the plant. Compatible with most other pesticides. The di-thio analog of methyl demeton. Low bee toxicity.

RELATED MIXTURE:

Ekamos, Ekatin-WF: — A Mixture of Parathion and Thiometon, developed by Sandoz Ltd. for use on cotton and other crops outside the U.S., especially for the control of whiteflies.

NAMES

PHORATE, AGRIMET, GEOMET, GRANUTOX, RAMPART, THIMET, TIMET

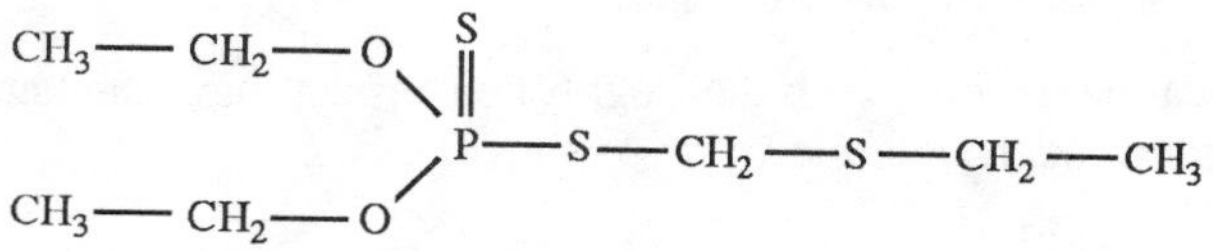

$$(CH_3-CH_2-O)_2P(=S)-S-CH_2-S-CH_2-CH_3$$

O,O-Diethyl-S-((ethylthio)methyl) phosphorodithioate

TYPE: Thimet is an organic phosphate soil insecticide-acaricide with considerable contact, systemic, and fumigant activity.

ORIGIN: American Cyanamid Company, 1954.

TOXICITY: LD_{50}-2 mg/kg. Rapidly absorbed through the skin and eyes.

FORMULATIONS: 10%, 15%, and 20% granules, 6 EC, 8 EC. Also sold combined with fertilizers and fungicides.

PHYTOTOXICITY: Treated plants are apparently stimulated. Slight chlorosis has been reported on some seedling plants, but they soon outgrow it. Injury may occur on tobacco and apples. Injury may occur if the granules come in direct contact with the seed of beans, corn, lettuce, tomatoes, and sugar beets.

USES: Alfalfa, barley, beans, corn, cotton, hops, lettuce, peanuts, potatoes, rice, sorghum, soybeans, sugarcane, sugar beets, tomatoes, and ornamentals. Widely used outside the U.S.

IMPORTANT PESTS CONTROLLED: Mites, aphids, greenbugs, thrips, leafhoppers, sorghum shootfly, leaf miners, corn rootworms, psyllids, cutworms, Hessian fly, foliar nematodes, wireworms, flea beetles, whiteflies, and many others.

RATES: Applied at 1/2-3 lb actual/A. Used at about 8 oz actual/100 lb seed treatment.

APPLICATION:

1. Soil Application
   a. Band Treatment-Distribute granules on each side of the row at planting time. Do not place in direct contact with the seed. Roots will absorb the material, translocating it into the leaves. Use high rates on heavier soils.
   b. Side-dressing-Incorporate into the soil as a side-dress application deep enough not to be disturbed by future cultivations. Irrigate as soon as possible.
2. Foliage treatments-Apply granular material when plants are dry in order to direct the granular material into the plant crowns. Spray when insects first appear, repeating as necessary.
3. Seed treatment-Coats seeds thoroughly before planting. A sticker may be added to increase the adhesion.

PRECAUTIONS: Do not graze livestock on treated crops. Keep treated seed out of reach of animals. Toxic to bees. Keep away from open flames. Non-compatible with alkaline compounds. Use on ornamentals should be by professional nurserymen only.

ADDITIONAL INFORMATION: Protects plants for 4-12 weeks after treatment. Compatibility is no problem because it is usually applied as a granular material. Soil mixed with PCNB as a cotton seed treatment for insect and fungi control. Used in federal, state, and industrial pine-seed orchards for pine-tip moth control.

NAMES

TERBUFOS, COUNTER

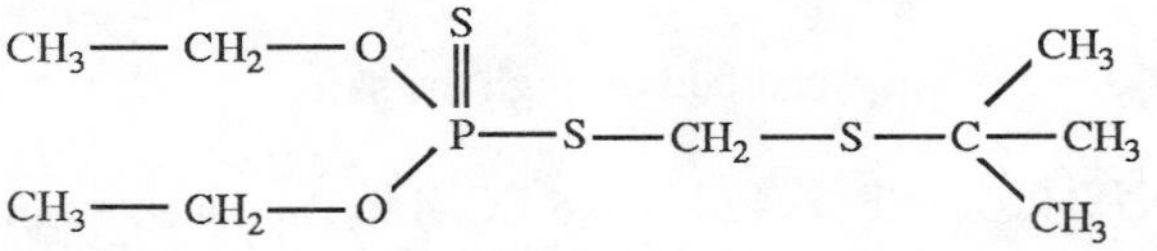

$$CH_3-CH_2-O \quad S \quad CH_3$$
$$P-S-CH_2-S-C-CH_3$$
$$CH_3-CH_2-O \quad CH_3$$

S[[(1,1-dimethylethyl) thio]methyl] O,O-diethylphosphorodithioate

TYPE: Counter is an organic phosphate compound used as a soil insecticide.

ORIGIN: American Cyanamid, 1973.

TOXICITY: LD_{50}-4.5 mg/kg. Avoid contact with eyes or skin.

FORMULATION: 15% granules.

PHYTOTOXICITY: Non-phytotoxic when used as directed.

USES: Corn, sorghum, and sugar beets. Used on these and other crops and bananas outside the U.S.

IMPORTANT PESTS CONTROLLED: Corn rootworms, wireworms, billbugs, seed corn maggots, seed corn beetles, white grubs, and other soil insects.

APPLICATION: Applied at planting time in a band, or applied directly to the seed furrow.

PRECAUTIONS: Toxic to fish and wild life.

NAMES

CARBOPHENOTHION, DAGADIP, GARRATHION, NEPHOCARP, THRITHION

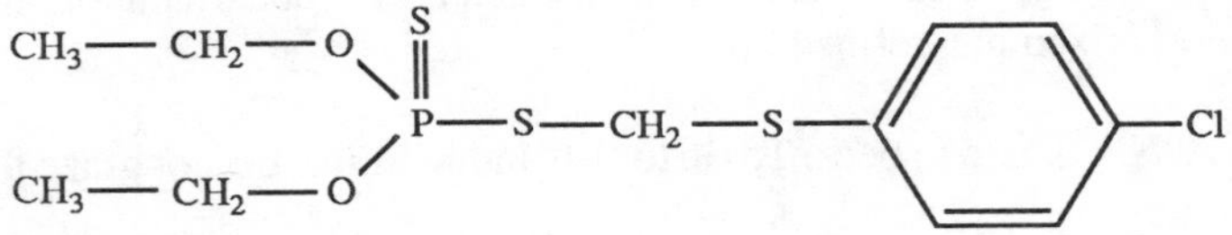

$$CH_3-CH_2-O \quad S$$
$$P-S-CH_2-S-C_6H_4-Cl$$
$$CH_3-CH_2-O$$

S-((p-Chlorophenylthio)methyl) O,O-diethyl phosphorodithioate

TYPE: Trithion is a nonsystemic, contact, and stomach-poison, organic phosphate insecticide-acaricide-ovicide.

ORIGIN: ICI Americas, 1955.

TOXICITY: LD_{50}-30 mg/kg. Absorbed through the skin.

FORMULATIONS: 25% WP, 2, 4, 6, and 8 EC, 4 lb/gal aqueous emulsion, 1-3% dusts.

PHYTOTOXICITY: No injury has been reported from dormant oil sprays. A 10-day waiting period between the application of dinitro compounds and Trithion is recommended because of the possibility of burning. Do not use on Charbono, Gamay, Golden Muscat, Muscat, or Italia grapes. Injury has been reported on grapefruit, kumquat, and citron. Crassulaceae, ferns, and a few sensitive rose varieties may be injured.

USES: Alfalfa, almonds, apples, apricots, beans, beets, cantaloupes, cherries, citrus, clover, corn, cotton, crab apples, cucumbers, eggplant, figs, grapes, nectarines, olives, onions, peaches, pears, peas, pecans, peppers, pimientos, plums, prunes, quinces, sorghum, soybeans, spinach, squash, strawberries, sugar beets, turf, tomatoes, walnuts, watermelons, and ornamentals. Used on beef cattle for the control of ectoparasites outside of the U.S.

IMPORTANT PESTS CONTROLLED: Mites, corn rootworm, aphids, leafhoppers, fleabeetles, psylla, scales, lice, lygus bugs, beetles, moths, butterflies, and livestock pests such as mites, hornflies, lice, bot tick, blue tick, brown tick, and others.

RATES: Applied at 1/4-5 lb/100 gal of water or 1/2-5 lb actual/A. Applied as a .1% solution to livestock.

APPLICATION:

1. Foliage-Apply with common application equipment at a uniform rate. Repeat as necessary.
2. Livestock-Used as a spray or dip on cattle. Dip vats and spray tanks should be thoroughly agitated while mixing. Caution should be taken in dipping calves under 3 months of age. It can also be used in backrubbers saturated with a 2% solution mixed in a refined oil.

PRECAUTIONS: Toxic to bees. Toxic to fish and wildlife. Do not graze the treated area.

ADDITIONAL INFORMATION: Long residual activity. Effective against eggs, as well as the postembryonic forms of many species. Combined with oil for dormant sprays. No off-flavor has resulted in the harvested crops. Compatible with most spray materials, including lime sulfur. Garrathion is the trade name of Trithion insecticide formulations used on livestock outside of the U.S.

NAMES

PROPYL THIOPYROPHOSPHATE, ASPON, **NPD**

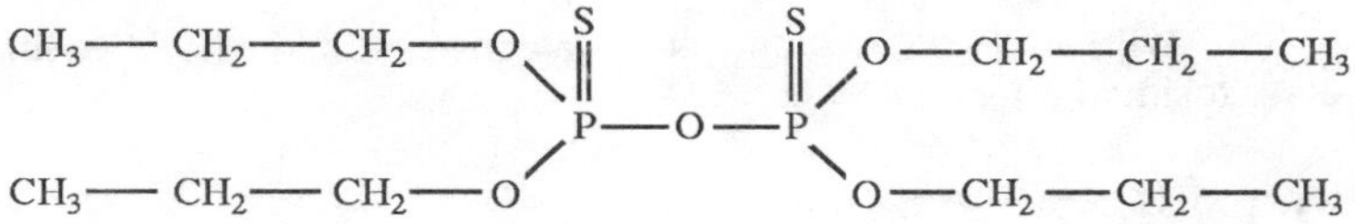

$$\begin{array}{ccccc} CH_3-CH_2-CH_2-O & S & & S & O-CH_2-CH_2-CH_3 \\ \diagdown & \| & & \| & \diagup \\ & P & -O- & P & \\ \diagup & & & & \diagdown \\ CH_3-CH_2-CH_2-O & & & & O-CH_2-CH_2-CH_3 \end{array}$$

O,O,O,O-Tetra-n-propyl dithiopyrophosphate

TYPE: Aspon is an organic phosphate insecticide, killing upon contact, and as a stomach-poison, with long residual effects.

ORIGIN: ICI Americas, 1963.

TOXICITY: LD_{50}-891 mg/kg.

FORMULATIONS: 6 EC, 5% granules.

USES: Turf.

IMPORTANT PESTS CONTROLLED: Chinch bugs.

RATES: Used at 1/4-1/2 lb actual/2000 sq ft of turf.

APPLICATION: Wet turf thoroughly and either spread on or spray the material. If sprayed on, use 100-150 gal of water/4000 sq ft. Irrigate immediately, washing chemical into the soil. Repeat at 6-week intervals.

PRECAUTIONS: Do not allow children or pets on treated area until the material has been washed into the soil. Do not contaminate streams, lakes, or ponds. Avoid drift. Do not use on any food or forage crops. Corrosive to steel. Toxic to fish and wildlife.

ADDITIONAL INFORMATION: A high initial kill is obtained. usually 95% kill within 48 hours can be expected. Control lasts 60-90 days.

NAMES

ETHION, DIETHION, EMBATHION, ETHANOX, ETHIOL, ETHODAN, HYLEMOX, ITOPAZ, NIALATE, RHODOCIDE

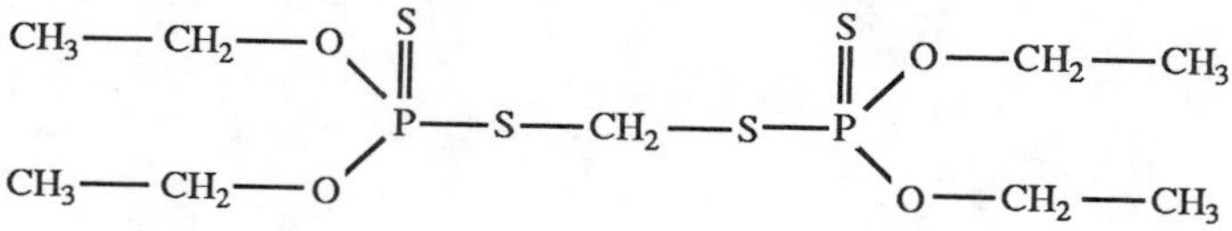

```
                      S                  S
CH3—CH2—O             ‖                  ‖  O—CH2—CH3
          \           ‖                  ‖ /
           P—S—CH2—S—P
          /                               \
CH3—CH2—O                                  O—CH2—CH3
```

O,O,O',O'-Tetraethyl S,S'-methylene bisphosphorodithioate

TYPE: Ethion is nonsystemic, organic phosphate insecticide-acaricide with relatively long residual effects.

ORIGIN: FMC Corp., 1959.

TOXICITY: LD_{50}-70 mg/kg. May be absorbed through the skin.

FORMULATIONS: 4 EC, 8 EC, 25% WP, 4% dust, 5% granules, plus various oil formulations and formulations with other insecticides.

PHYTOTOXICITY: Of the crops for which ethion uses have been accepted, the only ones upon which injury has been reported are certain varieties of apples. Therefore, do not use on Astrachen, Crimson, Davey, Duchess, Margaret Pratt, Melba, N.J. No. 3 (Brite-Mac), Wealthy, Williams Red (Red Williams), Yellow Transparent, or any apple variety maturing with, or before, Early McIntosh, as injury is likely to occur.

USES: Almonds, apples, apricots, beans, cherries, chestnuts, citrus, corn, cotton, cucumbers, eggplant, filberts, garlic, grapes, melons, nectarines, onions, peaches, pears, pecans, peppers, plums, prunes, sorghum, squash, strawberries, tomatoes, turf, walnuts, watermelons, and ornamentals.

IMPORTANT PESTS CONTROLLED: Mites, scale, aphids, codling moths, lygus bugs, thrips, leafhoppers, armyworms, leaf miners, pear psylla, Mexican bean beetles, seed maggots, and many others. Also controls the overwintering eggs of many species.

RATES: Applied at 1/4-1 lb actual/100 gal of water or 45 lb actual/A.

APPLICATION: Apply as either a dormant or foliar spray. Thorough coverage is necessary. Granules may be used in the furrow at onion planting time for control of the onion maggot.

PRECAUTIONS: Do not feed treated crop residue to livestock. Toxic to fish, bees, and wildlife. Do not combine with lime or copper, except in the oil formulations, or with lime or zinc sulfate plus lime. Do not store liquid formulations below 0°F.

NAMES

BIOTHION, TEMEPHOS, ABAT, ABATE, ABATHION, DIFENTHOS, NIMITEX, SWEBATE, TEMEPHOS

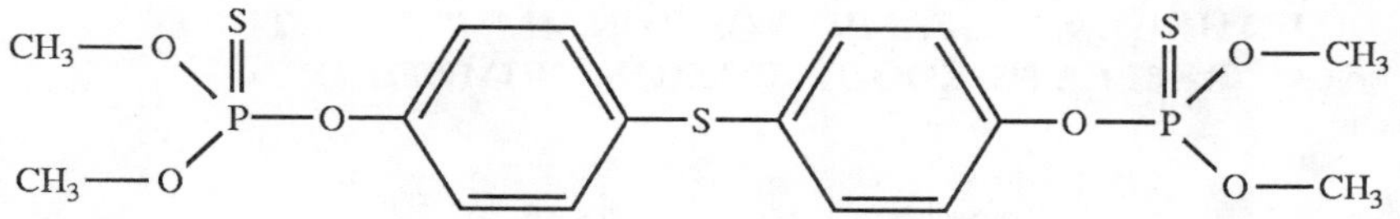

O,O,O',O'-Tetramethyl O,O-thiodi-_p_-phenylene phosphorothioate

TYPE: Abate is a nonsystemic, organic phosphate compound used as a selective insecticide with a long residual activity.

ORIGIN: American Cyanamid Company, 1964.

TOXICITY: LD_{50}-2030 mg/kg.

FORMULATIONS: 1% granules, 4 EC, 50% dispersible powders.

PHYTOTOXICITY: Non-phytotoxic when used at the recommended rates.

USES: Mosquito larvae and midge larvae control in ponds, lakes, marshes, swamps, etc. Widely used outside the U.S.

IMPORTANT PESTS CONTROLLED: Larvae of mosquitoes, midges, citrus thrips, gnats, punkies, sandflies, and thrips.

RATES: Used at 1/4-1 lb actual/A or 1/4-1 lb actual/100 gal of water. For mosquito larvae control use at .02-.1 lb actual/A.

APPLICATION: For control of mosquito larvae, apply as a light spray to the surface of ponds, marshes, swamps, and tidal areas. Also apply to the ground around the edges of such areas.

PRECAUTIONS: Young shrimp or crabs in treated tidal waters may be injured or killed. Toxic to fish and birds. Toxic to bees. Do not use on pasture crops used for forage.

ADDITIONAL INFORMATION: Relatively long residual qualities. Sometimes mixed with other compounds for broader-spectrum insect control. A highly selective compound. Most effective on the mosquito larvae, rather than the adults. Acceptable for use in potable water. Widely used in mosquito abatement programs. Used mainly in public health programs. Used in the control of human body lice.

NAMES

***DIMETHOATE,* CYGON, DAPHENE, DE-FEND, DEMOS-L40, DIMETHOGEN, FOSFAMID, FOSTION MM, PERFEKTHION, REBELATE, ROGOR, ROXION, TRIMETHION**

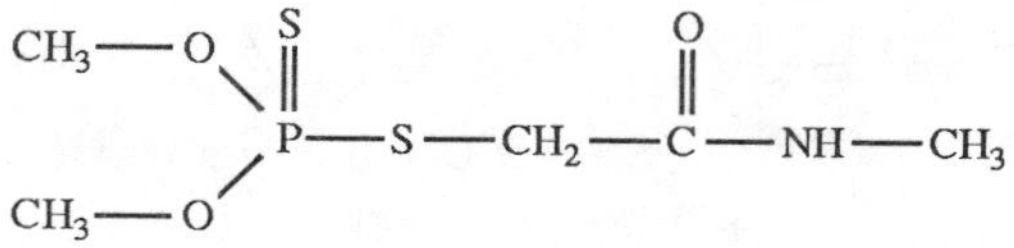

$$\begin{matrix} CH_3-O & S & & & O & & \\ \quad\diagdown & \| & & & \| & & \\ & P - S - & CH_2 - & & C - & NH - & CH_3 \\ \quad\diagup & & & & & & \\ CH_3-O & & & & & & \end{matrix}$$

O,O-Dimethyl S-(N-methylcarbamoylmethyl) phosphorodithioate

TYPE: Dimethoate is an organic phosphate insecticide-acaricide which shows both contact, residual, and systemic activity.

ORIGIN: American Cyanamid Co., 1956. Montecatini of Italy, BASF, Celamerck, and others are also producers of this product.

TOXICITY: LD_{50}-225 mg/kg.

FORMULATIONS: 2 and 2.67 EC, 4, 5, and 10% dusts, 25 and 50% WP, 5 and 10% granules.

PHYTOTOXICITY: Injury has been reported on certain varieties of walnuts, pines, peaches, hops, lemons, olives, figs, tomatoes, cotton, and beans. Russeting appears on Red and Golden Delicious apples and certain ornamentals. Do not treat seedling citrus.

USES: Alfalfa, apples, beans, broccoli, cabbage, cauliflower, citrus, collards, cotton, endive, grapes, kale, lettuce, melons, pears, peas, peppers, pecans, potatoes, safflower, soybeans, spinach, tobacco, tomatoes, turnips, wheat, ornamentals, Swiss chard, and agricultural premises. Outside the U.S., it is also used on cocoa, coffee, hops, tea, jute, sugarcane, sunflower, peanuts, artichokes, bananas, avocados, mangoes, olives, and other crops.

IMPORTANT PESTS CONTROLLED: Aphids, mites, codling moths, grasshoppers, plum curculio, pear psylla, scale, leafhoppers, thrips, loopers, drosophila, lygus bugs, leaf miners, flies, olive flies, whiteflies, Hassids, houseflies, and many others.

RATES: Applied at 1/16-4 lb actual/100 gal of water of 1/4-8 lb actual/A.

APPLICATION:

1. Granules—Place 5 inches away from row and 2 inches deep at time of planting. Water immediately.
2. Foliage — Apply spray or dust at a uniform rate with common application equipment. Repeat as necessary.
3. Households and barns — A residual wall spray for fly control.
4. Soil drench-Used on 1-3-year-old fruit trees. Apply in the furrow or basin around the base of the trees.
   Apply when insect injury to new growth appears.

PRECAUTIONS: Avoid excessive bee killing by not spraying a flowering crop. Use on agricultural premises only after animals have been removed. Do not use on greenhouse ornamentals.

ADDITIONAL INFORMATION: Effective against many insect larvae. Compatible with insecticides and fungicides that are not alkaline in reaction. Moves throughout the plant rapidly. Relatively slow acting against houseflies, but with up to 8 weeks' residual control. Half-life in the soil is 2-4 days, so it does not accumulate. Used to control fly maggots in manure piles.

NAMES

MECARBAM, AFOS, MURFOTOX, MUROTOX, PESTAN

$$(CH_3-CH_2-O)_2P(=S)-S-CH_2-C(=O)-N(CH_3)-C(=O)-O-CH_2-CH_3$$

O,O-Diethyl S-(N-ethoxycarbonyl N-methylcarbamoylmethyl) phosphorodithioate

TYPE: Mercarbam is an organic phosphate insecticide-acaricide with long residual activity.

ORIGIN: Murphy Chemical Company, Ltd., in England, 1958. Developed in Japan by Takeda Chemical Company. Marketed in France by Dow Chemical Co.

TOXICITY: LD_{50}-36 mg/kg.

FORMULATIONS: 25% WP, 80% liquid, 1.5-2% dusts, 40% and 68% EC.

PHYTOTOXICITY: Do not use on eggplant, as extreme injury may result.

USES: Not sold in the U.S. Used on apples as a preblossom treatment in Europe. Also used on cole crops, carrots, pears, cherries, peaches, plums, grains, citrus, and others outside the U.S.

IMPORTANT PESTS CONTROLLED: Aphids, applesuckers, pearsuckers, cabbage root fly, carrot fly, leafhoppers, mealybugs, scale, thrip, leaf miners, codling moths, and others.

RATE: Applied at 1-2 lb actual/A.

APPLICATION: Apply as a preblossom treatment, or after the tree has leafed out. Also applied as a soil insecticide.

PRECAUTIONS: Not for use in the U.S. Apply within 1 hour of mixing if applied with Bordeaux or lime sulfur. Keep livestock out of area for 1 week after application. Do not spray during blossom, or excessive bee kill may occur.

ADDITIONAL INFORMATION: Compatible with most other chemicals. 12 days-3 weeks' control. Considered a contact and stomach-poison. Nonsystemic. Possesses ovicidal properties.

NAMES

***PROTHOATE*, FAC, FOSTION, OLEOFAC, TELEFOS**

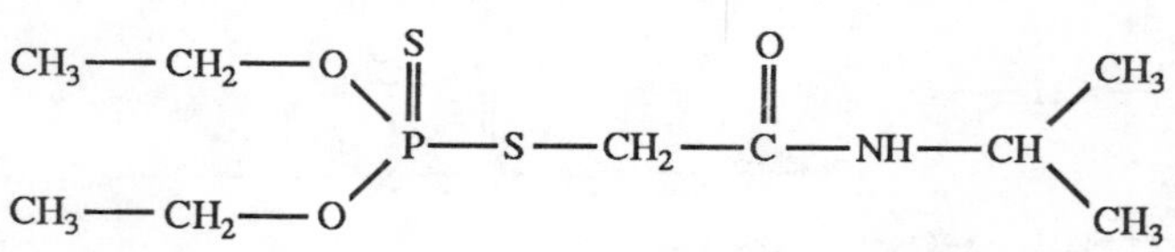

```
                       S                      O
CH3—CH2—O              ‖                      ‖                CH3
          \            ‖                      ‖               /
           P—S—CH2—C—NH—CH
          /                                                    \
CH3—CH2—O                                                      CH3
```

O,O-Diethyl S-(N-isopropylcarbamoylmethyl) phosphorodithioate

TYPE: FAC is an organic phosphate compound used as a contact insecticide-acaricide.

ORIGIN: Montecatini of Italy, 1956.

TOXICITY: LD_{50}-8 mg/kg.

FORMULATIONS: 20% and 40% EC, 3% dust, 5% granules, 40% WP.

USES: Used outside the U.S. on fruit trees, cotton, mangoes, ornamentals, sugar beets, sugarcane, tobacco, grapes, citrus, and vegetable crops.

PHYTOTOXICITY: Non-phytotoxic.

IMPORTANT PESTS CONTROLLED: Mites, thrips, aphids, lace bugs, and many others.

RATES: Applied at 1-2 lb actual/100 gal of water.

APPLICATION: Cover plant evenly and thoroughly. Repeat as necessary. Start application in the spring as soon as the first mites and insects are seen attacking the crop. Also used as a soil insecticide.

PRECAUTIONS: Do not mix with highly alkaline compounds. Not to be used in the U.S. Toxic to fish.

ADDITIONAL INFORMATION: Possesses ovicidal activity. Highly effective against sucking insects. Compatible with other pesticides. Related to dimethoate. Some systemic activity.

NAMES

CHLORMEPHOS, DOTAN

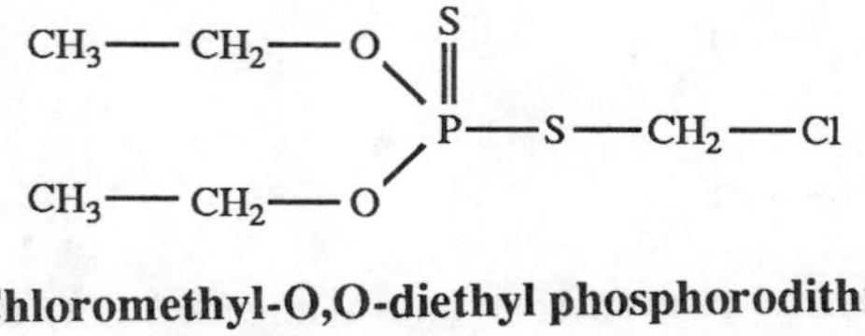

$$(CH_3—CH_2—O)_2P(=S)—S—CH_2—Cl$$

S-Chloromethyl-O,O-diethyl phosphorodithioate

TYPE: Chlormephos is an organic phosphorus insecticide with medium-long persistence that kills by contact.

ORIGIN: Murphy Chemical Co., 1968. Being developed and marketed by Rhone-Poulenc of France.

TOXICITY: LD_{50}-7 mg/kg.

FORMULATIONS: 5% and 10% granules.

PHYTOTOXICITY: None at normal rates. Soybeans and sorghum have been injured.

USES: A soil insecticide for cereals, potatoes, sugar beets, corn, tobacco, sugarcane, and other crops. Used outside the U.S.

IMPORTANT PESTS CONTROLLED: Wireworms, white grubs, houseflies, flea beetles, millipedes, symphilids, Hessian fly, leather jackets, root flies, crickets, and some other soil pests.

RATES: Applied at 400 g a.i./ha.

APPLICATION: Presowing or planting, with light incorporation, 1 inch deep into the soil. Banded application can also be made.

PRECAUTIONS: Toxic to fish. Avoid contact with skin or mouth. Not used in the U.S. Corrosive to some metals.

ADDITIONAL INFORMATION: Being marketed for use in Europe and Africa at the present time. Particularly effective against insects or larvae which live in the soil. Very rapid action, because the insects die as soon as they enter the treated area. Activity can last for 3 months.

NAMES

***FORMOTHION*, AFLIX, ANTHIO**

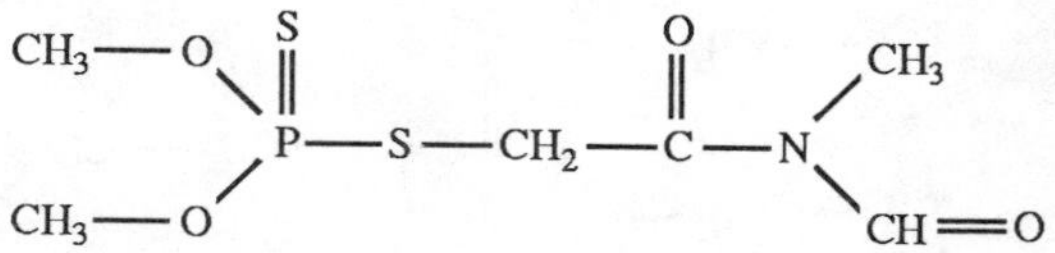

$$(CH_3-O)_2P(=S)-S-CH_2-C(=O)-N(CH_3)-CH=O$$

S-(N-formyl-N-methylcarbamoylmethyl) O,O-dimethyl phosphorodithioate

TYPE: Formothion is an organic phosphate insecticide-acaricide that possesses both contact and systemic activity.

ORIGIN: Sandoz Ltd. of Switzerland, 1959.

TOXICITY: LD_{50}-365-500 mg/kg.

FORMULATIONS: 25% and 33% EC, 35% ULV.

PHYTOTOXICITY: Generally considered Non-phytotoxic.

USES: Used outside the U.S. as a foliar spray or soil drench on a wide range of crops and ornamentals.

IMPORTANT PESTS CONTROLLED: Mites, aphids, psyllids, mealybugs, scales, whiteflies, thrips, codling moths, leaf miners, flies, and many others.

RATES: Applied at .3-.5 lb actual/A or at a .05% concentration.

APPLICATION: Applied at any time during the growing period when the pest population gets above the critical level. Cover the foliage thoroughly and evenly. Repeat when necessary. A soil application is made to hops, pouring the solution around the base of the plants.

PRECAUTIONS: Do not use in the U.S. Toxic to bees and fish. Do not mix with alkaline materials.

ADDITIONAL INFORMATION: Compatible with most pesticides. Long residual activity due to systemic action. Does not stain. Noncorrosive. Control for a 10-20 day period can be expected.

NAMES

***PHENTHOATE*, CIDIAL, DIMEPHENTHOATE, DIMETHENTHOATE, ELSAN, PAPTHION, ROGODIAL, TANONE**

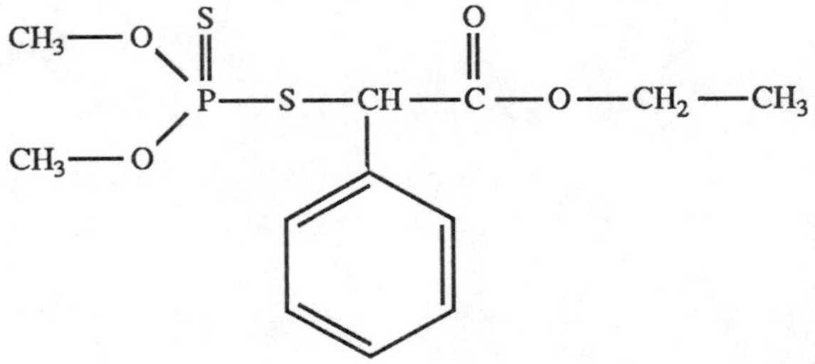

O,O-Dimethyl S-(alpha-ethoxycarbonylbenzyl)- phosphorodithioate

TYPE: Phenthoate is an organic phosphate compound used as a contact and stomach-poison insecticide-acaricide.

ORIGIN: Montedison of Italy, 1961. Sumitomo Chemical Co. Ltd. and Nissan Chemical Industries, of Japan also market this material.

TOXICITY: LD_{50}-300 mg/kg.

FORMULATIONS: 50 and 60% EC, 2% granules, 3% dust, 40% WP, 85% ULV, 75% ULV, 90% ULV, 4 EC.

PHYTOTOXICITY: Injury has resulted on some varieties of grapes, peaches, figs, and red-skinned apples.

USES: Being used outside the U.S. on rice, cereals, corn, citrus, coffee, cotton, sunflower, sugarcane, vegetables, fruit crops, and tea. Also used for mosquito control.

IMPORTANT PESTS CONTROLLED: Scales, bollworms, mealybugs, aphids, mosquitoes, mites, psyllids, borers, codling moths, leafhoppers, leaf miners, whiteflies, fleabeetles, cabbageworms, grape moths, mites, thrips, and many others.

RATES: Apply at .5-3 kg a.i./kg.

APPLICATION: Apply with common application equipment at the desired rates. Cover thoroughly and repeat as necessary.

PRECAUTIONS: Not registered for any usage in the U.S. Do not mix with alkaline materials. Do not apply to crops in full bloom in order to avoid bee injury. Toxic to fish.

ADDITIONAL INFORMATION: Nonsystemic. May be applied by air. Compatible with most other pesticides, except those alkaline in reaction. Excellent ovicidal activity.

NAMES

DIOXATHION, DELNAV, DELTIC, NAVADEL, RUPHOS

```
                              S
                              ||  O—CH2—CH3
                              || /
                          S—P
          O              /     \
  CH2 /      \ CH       /        O—CH2—CH3
   |           |
   |           |
  CH2 \      / CH                O—CH2—CH3
          O         \           /
                       S—P
                          ||  \
                          S    O—CH2—CH3
```

2,3-p-Dioxanedithiol-S, S-bis (O,O-diethyl phosphorodithioate)

TYPE: Delnav is an organic phosphate acaricide-insecticide showing stomach-poison, as well as a contact action, with long-lasting residual effects.

ORIGIN: Hercules Inc. (Now Nor-Am Chem. Co.), 1955.

TOXICITY: LD_{50}-40 mg/kg. Readily absorbed through the skin.

FORMULATIONS: 4 EC, 8 EC, cattle dips 15-30%.

PHYTOTOXICITY: Considered to be Non-phytotoxic when used at the recommended rates.

USES: Apples, apricots, cherries, citrus, grapes, peaches, pears, plums, prunes, quinces, walnuts, ornamentals, beef cattle, sheep, goats, horses, swine, and domestic pets, and as an industrial residual insecticide.

IMPORTANT PESTS CONTROLLED: Mites, ants, chiggers, crickets, spiders, thrips, apple maggot, codling moths, grape leafhoppers, ticks, fleas, flies, lice, and any others.

RATES: Applied at 1/4-1 lb actual/100 gal of water or 2 to 10 lb actual/A depending on the crop.

APPLICATION:

1. Foliage-Apply at a uniform rate with common application equipment. A second application may be made a week later, if needed.
2. Livestock-Apply as a dip, spray, pour-on treatment, or with backrubbers at any time before slaughter. Do not apply to dairy animals, or animals less than 3 months old.

PRECAUTIONS: Do not graze livestock on treated areas. Do not make reapplication within 3 months of first application to citrus if fruit is on the tree. Do not handle treated ornamentals for 24 hours after treatment. Toxic to fish and wildlife.

ADDITIONAL INFORMATION: Relatively harmless to bees. Some ovicidal activity has been shown. Residual lasts several weeks to several month, but has not shown direct systemic activity. Slow acting, since it takes 3-7 days for complete control. Compatible with insecticides and fungicides, except alkaline materials.

NAMES

***PHOSMET*, APPA, IMIDAN, INOVAT, KEMOLATE, PHTHALOPHOS, PMP, PROLATE**

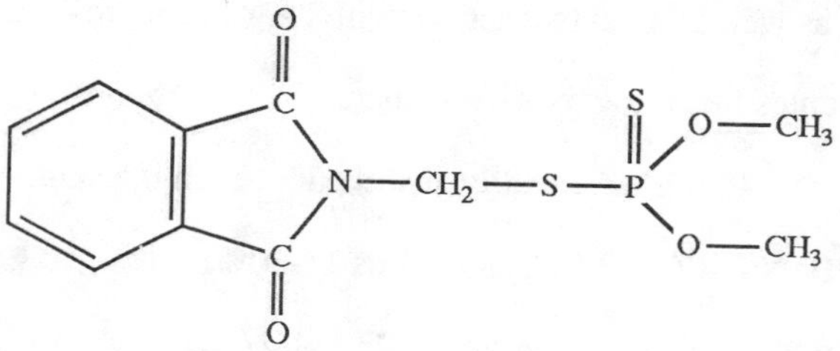

N-(Mercaptomethyl) phthalimide S-(O,O-dimethyl phosphorodithioate)

TYPE: Imidan is an organic phosphate insecticide-acaricide that controls by contact activity.

ORIGIN: ICI Americas, 1966.

TOXICITY: LD_{50}-113 mg/kg. May cause eye irritation.

FORMULATION: 50% WP.

PHYTOTOXICITY: None reported when used as directed.

USES: Alfalfa, apples, almonds, apricots, blueberries, cattle, corn, cotton, pistachios, cranberries, kiwi, peas, nectarines, potatoes, cherries, grapes, ornamentals, tomatoes, peaches, sweet potatoes, plums, prunes, pears, and beef cattle.

IMPORTANT PESTS CONTROLLED: Pear psylla, boll weevils, peach twig borer, cattle grubs, horn flies, Colorado potato beetle, leafhoppers, flea beetles,

Oriental fruitmoth, alfalfa weevil, aphids, codling moth, plum curculio, scales, and many others.

RATES: Applied at 1/2-3/4 actual/100 gal of water of 2 lb actual/A.

APPLICATION: Apply with common application equipment to give uniform coverage. Repeat as necessary. On livestock, used as either a pour-on treatment, or as a spray, or in backrubbers, or in a dip vat.

PRECAUTIONS: Do not treat young, stressed, or sick animals. Avoid drift. Do not apply when rain is expected, or before leaf surfaces are dry. Do not combine with alkaline material (lime sulfur, etc.). Do not treat dairy animals. Do not graze livestock in treated orchards. Toxic to fish and wildlife. Toxic to bees. Do not store at above 45°C.

ADDITIONAL INFORMATION: Suppresses mites when used in a regular spray program. Compatible with most other pesticides. Used on sweet potatoes in storage to control the sweet potato beetle.

NAMES

***METHIDATHION*, SUPRACIDE, ULTRACIDE**

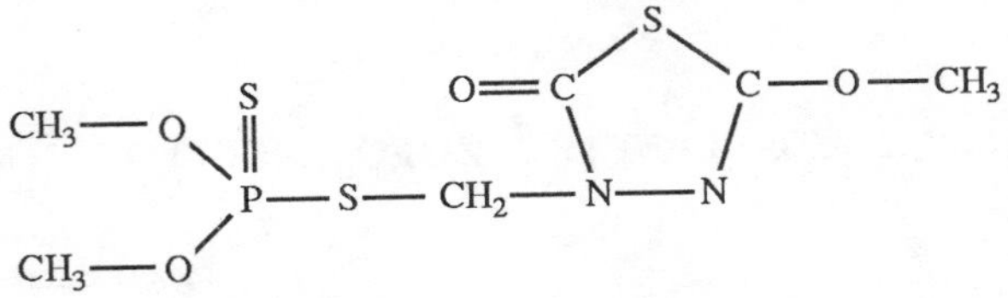

```
                                        S
                                      /   \
              S              O═C           C—O—CH3
CH3—O         ‖                 \         /
      \       ‖                  \       /
        P—S—CH2——————————N———N
      /
CH3—O
```

O,O-Dimethyl phosphorodithioate S-ester with 4-(mercaptomethyl)-2-methoxy-1,3,4-thiadiazolin-5-one

TYPE: Supracide is an organic phosphate compound being used as an insecticide-acaricide with relatively long residual effectiveness.

ORIGIN: CIBA-Geigy Corp., 1966.

TOXICITY: LD_{50}-25 mg/kg.

FORMULATIONS: 2 EC, 20% and 40% WP, 40% EC, ULV 250.

PHYTOTOXICITY: Non-phytotoxic at the recommended rates. Apples have had reported injury.

USES: Alfalfa, almonds, apples, apricots, artichokes, cherries, citrus, clover, cotton, grass, nectarines, olives, peaches, pears, pecans, plums, potatoes, prunes, safflower, sorghum, sunflower, tobacco, and walnuts. Being used in Europe on a number of crops.

IMPORTANT PESTS CONTROLLED: Codling moths, alfalfa, weevils, leafhoppers, spittlebugs, hornworms, budworms, fleabeetles, mites, scale, leafminers, boll worms, lygus bugs, aphids, potato beetles, cabbageworms, hornworms, plum curculio, thrips, boll weevils, and others.

RATES: Applied at 1/2-1 lb actual/A.

APPLICATION: Apply at 1/8-1/2 lb actual/100 gal of water or at 1/4-1 lb a.i./A. Apply evenly and thoroughly. Repeat as necessary.

PRECAUTIONS: Mixing with alkaline materials may reduce the effectiveness. Toxic to fish and wildlife. Toxic to bees, so do not apply during blooming period. Do not enter treated fields the same day as treatment.

ADDITIONAL INFORMATION: Residual effectiveness for 3-5 weeks has been noted. Compatible with a number of other pesticides. May be applied by air.

NAMES

PHOSALONE, AZONFENE, BENZOFOS, RUBITOX, ZOLONE

s-6-chloro-2,3-dihydro-2-oxobenzoxazol-3-yl methyl O,O-diethylphosphorodithioate

TYPE: Phosalone is a organic phosphate insecticide-acaricide with contact and stomach-poison activity.

ORIGIN: Rhone-Poulenc, 1963.

TOXICITY: LD_{50}-90 mg/kg. May cause skin irriatation.

FORMULATIONS: 3 EC, 25% WP.

PHYTOTOXICITY: Injury has occurred to a number of plants only when used at higher than the normal recommended rates. Golden Delicious apples as well as other yellow varieties, have been injured.

USES: Almonds, artichokes, potatoes, pecans, and walnut. Used outside the U.S. on these and many other crops.

IMPORTANT PESTS CONTROLLED: Mites, aphids, leafhoppers, peach twig borer, codling moths, pear psylla, plum curculio, and others.

RATES: Applied at 3-8 lb actual/A.

APPLICATION: Apply evenly and thoroughly in sufficient water to insure coverage. Repeat as necessary.

PRECAUTIONS: Do not allow livestock to graze in treated areas. Do not store below 32°F. Toxic to fish. Do not use with surfactants. Toxic to bees, so do not apply when they are foraging.

ADDITIONAL INFORMATION: A broad-spectrum pesticide with rapid killing ability. Approximately 12-20 days' control may be expected. Does not move in the soil. Degrades even more rapidly in the soil than parathion. Nonsystemic in action. Relatively safe to use around bees. Compatible with most other pesticides. Activity is not modified by low temperatures.

NAMES

***FONOFOS*, CAPFOS, DYFONATE**

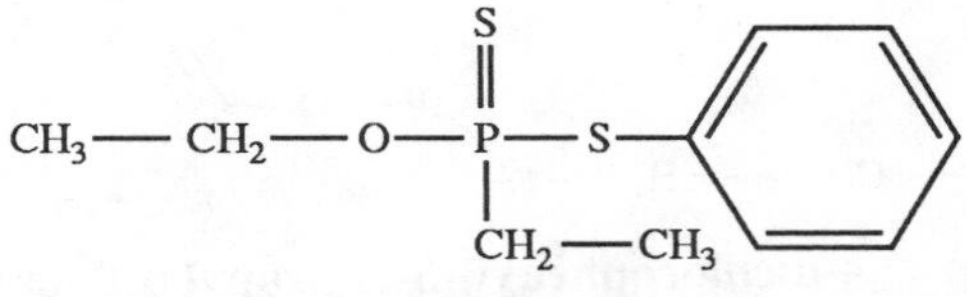

S
||
CH_3— CH_2—O—P—S—
|
CH_2—CH_3

O-Ethyl S-phenyl ethylphosphorodithioate

TYPE: Dyfonate is an organic phosphate compound used as a selective soil insecticide.

ORIGIN: ICI Americas, 1967.

TOXICITY: LD_{50}-8 mg/kg.

FORMULATIONS: 10% granules, 20% granules, 4 EC.

PHYTOTOXICITY: Non-phytotoxic when used as directed.

USES: Asparagus, beets, beans, mint, strawberries, sugarcane, broccoli, Brussels sprouts, cabbage, cauliflower, sorghum, corn, onions, peanuts, potatoes, radishes, sugar beets, sweet potatoes, tobacco, and turf.

IMPORTANT PESTS CONTROLLED: Garden symphylans, cabbage maggot, aphids, European corn borer, onions, maggot, corn rootworms, and wireworms.

RATES: Usually applied at 1/2-4 lb actual/A.

APPLICATION: Incorporate into the soil 2-3 inches deep prior to seeding or transplanting. Incorporation is usually done by disking or placed directly ahead of the press wheel.

PRECAUTIONS: Hazardous to wildlife and fish. Avoid drift. Birds feeding on treated areas may be killed. Do not rotate the treated crop with carrots.

ADDITIONAL INFORMATION: No fumigant activity. Contact kill only with residual activity, giving season-long control. May be used in wet or dry weather. May be mixed with fertilizer.

NAMES

PROTHIOPHOS, BIDERON, NTN-8629, TOKUTHION, TOYODAN, TOYOTHION

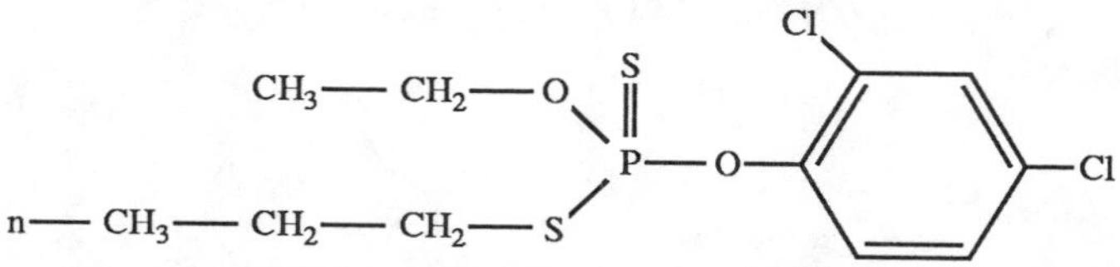

O-Ethyl-O-(2,4-dichlorophenyl)-S-N-propyl dithiophosphate

TYPE: Tokuthion is an organic phosphate compound used as a broad-spectrum contact and stomach-poison insecticide.

ORIGIN: Nihon Tokushu Noyaku Seizo KK of Japan, 1969. Also being developed by Bayer AG of Germany.

TOXICITY: LD_{50}-925 mg/kg.

FORMULATIONS: 4 EC, 40% WP, 2 and 3% granules, 4% bait, .2% dust.

PHYTOTOXICITY: Russeting has been known to occur on Golden Delicious apples.

USAGE: Not for sale or use in the U.S. Used in other countries on apples, pears, grapes, citrus, vegetables, corn, potatoes, sugarcane, sugar beets, tea, tobacco, hops, ornamentals, and others.

IMPORTANT INSECTS CONTROLLED: Aphids, flies, mosquitoes, thrips, mealybugs, caterpillars, leaf rollers,wireworms, cutworms, termites, white grubs, mites, armyworms, diamondback moth, gypsy moth, European corn borers, and many others.

APPLICATION:

1. Foliage-Apply when insects appear, and repeat as necessary. Does not control leafhoppers or plant bugs.
2. Soil-Applied either to the soil or incorporated into the top 6 inches of soil at planting time.
3. Public health-Used on a weekly basis to control flies, mosquitoes in refuse dumps, lavatories, market places, house walls, poultry houses, etc.

PRECAUTIONS: Not for sale or use in the U.S. Does not control mining or boring insects. Toxic to fish. Do not mix with alkaline materials.

ADDITIONAL INFORMATION: Broad spectrum. Low toxicity to mammals, birds, and fish. Safe to bees if used at the recommended rates. No systemic activity. Relatively slow in its initial activity, but gives good residual in the plants and the soil. Compatible with other pesticides. Some activity against aphids.

NAMES

SULPROFOS, MERDAFOS, BAY-NTN-9306, BOLSTAR, HELOTHION

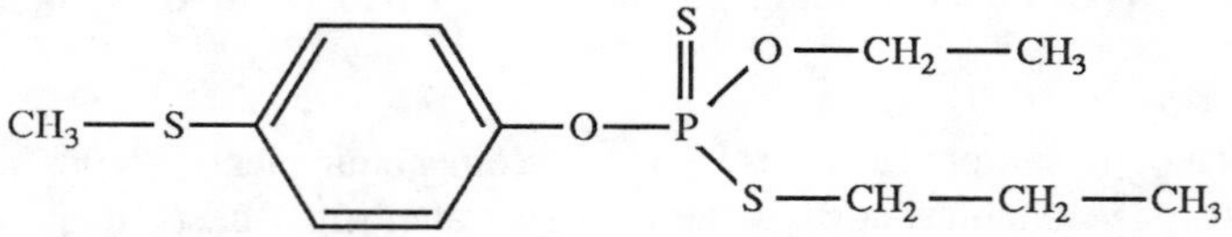

O-Ethyl O-[4-(methylthio) phenyl] S-propyl phosporodithioate

TYPE: BOLSTAR is an organic phosphate compound used as a contact, selective insecticide.

ORIGIN: Nitokuno of Japan, 1973. Being developed by Bayer AG of Germany, and in the U.S. by Mobay Chemical Co.

TOXICITY: LD_{50}-65 mg/kg.

FORMULATION: 6 EC.

PHYTOTOXICITY: Some injury has occurred to the foliage of cole crops, peanuts, peppers, and deciduous fruits. Numerous applications sometimes give cotton leaves a red discoloration.

USES: Cotton and soybeans. Experimentally on corn, tobacco, and certain vegetable crops. Outside the U.S. it is used on these and other crops. Experimentally it is being tested as a forest insecticide.

IMPORTANT PESTS CONTROLLED: Bollworms, tobacco budworms, mealybugs, leafhoppers, mites, whieflies, armyworms, and others.

RATES: Applied at 1/2-1.5 lb actual/A.

APPLICATION: Apply to the foliage when insect appear, and repeat as necessary.

PRECAUTIONS: Toxic to fish and wildlife.

ADDITIONAL INFORMATION: Particularly effective against Lepidopterous insects. Miticide activity has been shown. Appears to have poor soil stability. Most effective on the first to third instar larvae. Nonsystemic. May be mixed with certain other pesticides.

GLOSSARY

SPRAYER CALIBRATION

Calibration of your spraying equipment is very important. It should be done at least every other day of operation to insure application of the proper dosages. This is probably the most important step in your whole spraying operation since applying incorrect amounts may do much more damage than good.

If a lower rate is desired it may be obtained by increasing the speed, reducing the speed, increasing the pressure or changing to a larger nozzle or a combination of the three.

GENERAL CALIBRATION

I. **Method I.**
   A. Measure out 660 feet.
   B. Determine the amount of spray put out in traveling this distance at the desired speed.
   C. Use this formula:
   $$\text{gallons/acre} = \frac{\text{gallons used in 660 feet x 66}}{\text{swath width in feet}}$$
   * D. Fill the tank with the desired concentration.

II. **Method II.**
   A. Fill spray tank and spray a specified number of feet.
   B. After spraying refill tank measuring the quantity of material needed for refilling.
   C. Use this formula:
   $$\text{gallons/acre} = \frac{\text{43560 x gallons delivered}}{\text{swath lenght (ft.) x swath width (ft.)}}$$
   * D. Fill the tank with the desired concentrate.

III. **Method III.**
   A. Measure 163 feet in the field.
   B. Time tractor in 163 feet. Make two passes to check accuracy.
   C. At edge of field adjust pressure valve until you catch 2 pints (32 ounces) of spray in the same amount of time it took to run the 163 feet. Be sure tractor is at the same throttle setting. You are now applying 20 gallons per acre on a 20 inch nozzle boom spacing.
   D. For each inch of nozzle spacing on boom, increase time by 5% or reduce the volume by 5%.

---

* If you have calibrated your rig and it is putting out 37 gpa, the required dosage is 4 lbs. actual/acre. Therefore, for every 37 gallons of carrier (water, oil, etc.) in the spray tank you add 4 lbs. of active material.

USEFUL FORMULAE

1. To determine the amount of active ingredient needed to mix in the spray tank.
   No. of gallons or pounds =
   $$\frac{\text{No. of acres to be sprayed x pounds active ingredient required per A}}{\text{pounds active ingredient per gallon or per pound}}$$

2. To determine the amount of pesticide needed to mix a spray containing a certain percentage of the active ingredient.
   No. of gallons or pounds =
   $$\frac{\text{gallons of spray desired x \% active ingredient wanted x 8.345}}{\text{lbs. active ingredient per gallon or pound X 100}}$$

3. To determine the percent active ingredient in a spray mixture.
   Percent =
   $$\frac{\text{lbs. or gallons of concentrate used (not just active ingredient) x \% active ingredient in the concentrate}}{\text{gallons of spray X 8.345 (weight of water/gallon)}}$$

4. To determine the amount of pesticide needed to mix a dust with a given percent active ingredient.
   pounds material =
   $$\frac{\text{\% active ingredient wanted x lbs. of mixed dust wanted}}{\text{\% active ingredient in pesticide used}}$$

5. To determine the size of pump needed to apply a given number of gallons/acre.
   pump capacity =
   $$\frac{\text{gallons/acre desired x boom width (feet) x mph}}{495}$$

6. To determine the nozzle capacity in gallons per minute at a given rate/acre and miles/hour.
   Nozzle capacity =
   $$\frac{\text{gallons/acre x nozzle spacing (inches) x mph}}{5940}$$

7. To determine the acres per hours sprayed.
   Acres per hour =
   $$\frac{\text{swath width (inches) x mph}}{100}$$

8. To determine the rate of speed in miles per hour.
   1. Measure off a distance of 300 to 500 feet.
   2. Measure in seconds the time it takes the tractor to go the marked off distance.

3. Multiply .682 times the distance traveled in feet and divide product by the number of seconds.

$$\text{MPH} = \frac{.682 \times \text{distance}}{\text{seconds}}$$

9. To determine the nozzle flow rate.
Time the seconds necessary to fill a pint jar from a nozzle.
Divide the number of seconds into 7.5.

$$\text{gallons/minute/nozzle} = \frac{7.5}{\text{seconds}}$$

10. To determine the gallons per minute per boom.
Figure out the gallons/minute/nozzle and multiply by the number of nozzles.

11. To determine the gallons per acre delivered.

$$\frac{5940 \times \text{gallons/minute/nozzle} = \text{gpa}}{\text{nozzle spacing (inches)} \times \text{mph}}$$

12. To determine the acreage sprayed per hour.

$$\text{acres sprayed/hour} = \frac{\text{boom width (feet)} \times \text{mph}}{12}$$

This allows 30% of time for filling, turning, etc.

13. Sprayer Tank Capacity
Calculate as follows:

1. Cylindrical Tanks:
Multiply the length in inches times the square of the diameter in inches and multiply the product by .0034.
length x diameter squared x .0034 = number of gallons.

2. Elliptical Tanks:
Multiply the length in inches times the short diameter in inches times the long diameter in inches times .0034.
length x short diameter x long diameter x .0034 = number of gallon.

3. Rectangular Tanks:
Multiply the length times the width times the depth in inches and multiply the product by .004329.
length x width x depth x .004329 = number of gallons.

14. To determine the acres in a given area.
Multiply the length in feet times the width in feet times 23. Move the decimal point 6 places to the left to give the actual acres.

CONVERSION TABLES (U.S.)

Linear measure -

1 foot - 12 inches
1 yard - 3 feet
1 rod - 5.5 yards — 16.5 feet
1 mile - 320 rods — 1760 yards — 5280 feet

Square Measure -

1 square foot (sq. ft.) — 144 square inches (sq. inch)
1 square yard (sq. yd.) — 9 square feet
1 square rod (sq. rd.) — 272.25 sq. ft. — 30.25 sq. yd.
1 acre - 43560 sq. ft. — 4840 sq. yds. — 160 sq. rds.
1 square mile - 640 acres

Cubic Measure -

1 cubic foot (cu. ft.) — 1728 cubic inches (cu. in) 29.922 liquid quarts = 7.48 gallons
1 cubic yard - 27 cubic feet

Liquid Capacity Measure -

1 tablespoon - 3 teaspoons
1 fluid ounce - 2 tablespoons
1 cup - 8 fluid ounces
1 pint - 2 cups — 16 fluid ounces
1 quart - 2 pints — 32 fluid ounces
1 gallon - 4 quarts — 8 pints — 128 fluid ounces

Weight Measure -

1 pound (lb.) - 16 ounces
1 hundred weight (cwt.) - 100 pounds
1 ton - 20 cwt. — 2000 pounds

Rates of Application -

1 ounce/sq. ft. — 2722.5 lbs./acre
1 ounce/sq. yd. — 302.5 lbs./acre
1 ounce/100 sq. ft. — 27.2 lbs./acre
1 pound/100 sq. ft. — 435.6 lbs./acre
1 pound/1000 sq. ft. — 43.6 lbs./acre
1 gallon/acre — 3 ounce/1000 sq. ft.
5 gallons/acre — 1 pint/1000 sq. ft.
100 gallons/acre — 2.5 gallons/1000 sq. ft. — 1 quart/100 sq. ft.
100 lbs./acre — 2.5 lbs./1000 sq. ft.

Important Facts -

Volume of sphere — diameter ^{3} x .5236
Diameter — Circum. x .31831
Area of circle — dia. ^{2} x .7854
Area of ellipse — prod. of both dia. x .7854
Vol. of cone — area of base x 1/3 ht.
1 cu. ft. water = 7.5 gallons = 62.5 lbs.
Pressure in psi — ht. (ft) x .434
1 acre = 209 feet square
ppm = % x 10,000
% — ppm
10,000
1% by volume — 10,000 ppm

TABLE OF CONVERSION FACTORS

To Convert From	To	Multiply By
Cubic feet	gallons	7.48
Cubic feet	liters	28.3
Gallons	milliliters	3785
Grams	pounds	.0022
Grams/liter	parts/million	1000
Grams/liter	pounds/gallon	.00834
Liters	cubic feet	.0353
Milligrams/liter	parts/million	1
Milliliters/gallons	gallons	.0026
Ounces	grams	28.35
Parts/million	grams/liter	.001
Parts/million	pounds/million gallons	8.34
Pounds	grams	453.59
Pounds/gallon	grams/liter	111.83

1 gram = .035 ounce
1 kilogram = 2.2 lbs.
1 quintal = 100 kg. = 221 lbs.

1 metric ton = 1000 kg. = 2,205 lbs.
1 hectare = 2.5 acres
1 meter = 39.4 inches
1 kilometer = .6 mile

CONVERSION TABLE

1 kilogram (kg) = 1000 grams (g) = 2.2 lbs.
1 gram (g) = 1000 milligrams (mg) = .035 ounce
1 liter = 1000 milliliters (ml) or cubic centimeters (cc) = 1.058 quarts
1 milliliter or cubic centimeter = .034 fluid ounce
1 milliliter or cubic centimeter of water weighs 1 gram
1 liter of water weighs 1 kilogram
1 lb. = 453.6 grams
1 ounce = 28.35 grams
1 pt. of water weighs approximately 1 lb.
1 gallon of water weighs approximately 8.34 lbs.

1 gallon = 4 qts. = 3.785 liters
1 qt. = 2 pts. = .946 liters
1 pt. = .473 liters
1 fluid ounce = 29.6 milliliters or cubic centimeters

1 part per million (ppm) = 1 milligram/liter
= 1 milligram/kilogram
= .0001 percent
= .013 ounces by in 100 gallons of water

1 percent = 10.000 ppm
= 10 grams per liter
= 10 grams per kilogram
= 1.33 ounces by weight per gallon of water
= 8.34 pounds/100 gallons of water

.1 percent = 1000 ppm = 1000 milligrams/liter
.01 percent = 100 ppm = 100 milligrams/liter
.001 percent = 10 ppm = 10 milligrams/liter
.0001 percent = 1 ppm = 1 milligrams/liter

1 lb./acre = 1.12 kg/ha.

CHEMICAL ELEMENTS

Name	Symbol	Atomic Weight	Valance
Aluminum	Al	26.97	3
Antimony	Sb	121.76	3, 5
Arsenic	As	74.91	3, 5
Barium	Ba	137.36	2
Bismuth	Bi	209.00	3, 5
Boron	B	10.82	3, 0
Bromine	Br	79.916	1, 3, 5, 7
Cadmium	Cd	112.41	2
Calcium	Ca	40.08	2
Carbon	C	12.01	2, 4
Chlorine	Cl	35.457	1, 3, 5, 7
Cobalt	Co	58.94	2, 3
Copper	Cu	63.57	1, 2
Fluorine	F	19.00	1
Hydrogen	H	1.008	1
Iodine	I	126.92	1, 3, 5, 7
Iron	Fe	55.85	2, 3
Lead	Pb	207.21	2, 4
Magnesium	Mg	24.32	2
Mercury	Hg	200.61	1, 2
Molybdenum	Mo	95.95	3, 4, 6
Nickel	Ni	58.69	2, 3
Nitrogen	N	14.008	3, 5
Oxygen	O	16.00	2
Phosphorus	P	30.98	3, 5
Potassium	K	39.096	1
Selenium	Se	78.96	3
Silicon	Si	28.06	4
Silver	Ag	107.88	1
Sodium	Na	22.997	1
Sulfur	S	32.06	2, 4, 6
Thallium	Tl	204.29	1, 3
Tin	Sn	118.70	2, 4
Titanium	Ti	47.90	3, 4
Uranium	U	238.17	4, 6
Zinc	Zn	65.38	2

WIDTH OF AREA COVERED TO ACRES PER MILE TRAVELED

Width of Strip (feet)	Acres/mile
6	.72
10	1.21
12	1.45
16	1.93
18	2.18
20	2.42
25	3.02
30	3.63
50	6.04
75	9.06
100	12.1
150	18.14
200	24.2
300	36.3

TEMPERATURE CONVERSION TABLE RELATIONSHIP OF CENTIGRADE AND FAHRENHEIT SCALES

°C	°F	°C	°F
-40	-40	25	77
-35	-31	30	86
-30	-22	35	95
-25	-13	40	104
-20	-4	45	113
-15	5	50	122
-10	14	55	131
-5	23	60	140
0	32	80	176
5	41	100	212
10	50		
15	59		
20	68		

PROPORTIONATE AMOUNTS OF DRY MATERIALS

Water	Quantity of Material				
100 gallons	1 lb.	2 lbs.	3 lbs.	4 lbs.	5 lbs.
50 gallons	8 oz.	1 lb.	24 oz.	2 lbs.	2 1/2 lbs.
5 gallons	3 tbs.	1 1/2 oz.	2 1/2 oz.	3 1/4 oz.	4 oz.
1 gallon	2 tsp.	3 tsp.	1 1/2 tbs.	2 tbs.	3 tbs.

PROPORTIONATE AMOUNTS OF LIQUID MATERIALS

Water	Quantity of Material		
100 gallons	1 qt.	1 pt.	1 cup
50 gallons	1 pt.	1 cup	1/2 cup
5 gallons	3 tbs.	5 tsp.	2 1/2 tsp.
1 gallon	2 tsp.	1 tsp.	1/2 tsp.

MILES PER HOUR CONVERTED TO FEET PER MINUTE

MPH	fpm
1	88
2	176
3	264
4	352

EMULSIFIABLE CONCENTRATE PERCENT RATINGS IN POUNDS ACTUAL PER GALLON

%EC	lbs./gallon
10-12	1
15-20	1.5
25	2
40-50	4
60-65	6
70-75	8
80-100	10

CONVERSION TABLE FOR LIQUID FORMULATIONS

Concentration of Active Ingredient in Formulations, lbs./gal.

Rate Desired Lbs./A	1	2	2.5	3	4	5	6
			(ml of formulation per 100 square feet)				
1	8.67	4.33	3.47	2.89	2.17	1.73	1.44
2	17.3	8.67	6.93	5.78	4.33	3.47	2.89
3	26.0	13.0	10.4	8.67	6.50	5.20	4.33
4	34.8	17.4	13.9	11.6	8.69	6.95	5.79
5	43.4	21.7	17.4	14.5	10.0z	8.68	7.24
6	52.1	26.0	20.8	17.4	13.0	10.4	8.68
7	60.8	30.4	24.3	20.3	15.2	12.2	10.1
8	69.4	34.7	27.8	23.1	17.4	13.9	11.6
9	78.1	39.0	31.2	26.0	19.5	15.6	13.0
10	86.7	43.3	34.7	28.9	21.7	17.3	14.4

***Example:** To put out a 100 sq. ft. plot at the rate of 5 lbs./A active ingredient using a formulation containing 4 lbs./gal. active ingred., use 10.9 ml. of the 4 lbs./gal. form. & distribute evenly.

CONVERSION TABLE FOR GRANULAR FORMULATIONS

Concentration of Active Ingredient in Formulation

Rate Desired Lbs./A	20%	15%	10%	7.5%	5%	4%	3%	2%	1%
	(Grams of formulation per 100 square feet)								
1	5.2	6.94	10.4	13.86	20.8	26.0	34.66	52.0	104.0
2	10.4	13.9	20.8	27.7	41.7	52.0	69.3	104.0	208.0
3	15.6	20.8	31.2	41.6	62.5	78.0	103.9	156.0	312.0
4	20.8	27.8	41.7	55.4	83.3	104.0	138.6	208.0	416.0
5	26.0	34.7	52.1	69.3	104.0*	130.0	173.3	260.0	520.0
6	31.2	41.6	62.5	83.2	125.0	156.0	207.9	312.0	624.0
7	36.4	45.6	72.9	97.0	146.0	182.0	242.6	364.0	728.0
8	41.7	55.5	83.3	110.9	167.0	208.0	277.3	416.0	832.0
9	46.9	62.5	93.7	124.7	187.0	234.0	311.9	468.0	936.0
10	52.1	69.4	104.0	138.6	208.0	260.0	346.6	520.0	1040.0
15	78.0	104.1	156.0	207.9	312.0	390.0	519.2	780.0	1560.0
20	104.0	138.8	208.0	277.2	416.0	520.0	693.2	1040.0	2080.0
25	130.0	173.5	260.0	346.5	520.0	650.0	866.5	1300.0	2600.0
30	156.0	208.2	312.0	415.8	624.0	780.0	1039.8	1560.0	3120.0

***Example:** To put out a 100 sq. ft. plot at the rate of 5 lbs./A active ingredient using a formulation containing 5% active material, use 104.0 grams of the 5% formulation and distribute it evenly over the 100 sq. ft.

CONVERSION TABLE FOR DRY FORMULATIONS

Concentration of Active Ingredient in Formulation

Rate Desired Lbs./A	100%	90%	80%	75%	70%	60%	50%	40%	30%	25%	20%	10%	5%
	(Grams of formulation per 100 square feet)												
1	1.04	1.16	1.30	1.39	1.49	1.74	2.08	2.60	3.47	4.17	5.21	10.4	20.8
2	2.08	2.31	2.60	2.78	2.98	3.47	4.17	5.21	6.94	8.33	10.4	20.8	41.7
3	3.12	3.47	3.90	4.17	4.46	5.20	6.25	7.81	10.4	12.5	15.6	31.2	62.5
4	4.17	4.63	5.21	5.55	5.95	6.94	8.33*	10.4	13.9	16.7	20.8	41.7	83.3
5	5.21	5.79	6.51	6.94	7.44	8.68	10.4	13.0	17.4	20.8	26.0	52.1	104
6	6.25	6.94	7.81	8.33	8.93	10.4	12.5	15.6	20.8	25.0	31.2	62.5	125
7	7.29	8.10	9.11	9.72	10.4	12.1	14.6	18.2	24.3	29.2	36.4	72.9	146
8	8.33	9.26	10.4	11.1	11.9	13.9	16.7	20.8	27.8	33.3	41.7	83.3	167
9	9.37	10.4	11.7	12.5	13.4	15.6	18.7	23.4	31.2	37.5	46.9	93.7	187
10	10.4	11.6	13.0	13.9	14.9	17.4	20.8	26.0	34.7	41.7	52.1	104	208

***Example:** To put out a 100 sq. ft. plot at the rate of 4 lbs./A active ingredient using formulation containing 50% active ingredient, use 8.33 grams of the 50% formulation and distribute evenly over the 100 sq. ft.

DETERMINE THE NUMBER OF ROWS TO THE ACRE

Length of Rows

Rows/Acre	32"	36"	38"	40"	60"
1	16335	14520	13756	13068	8712
2	8168	7260	6878	6534	4356
3	5445	4840	4585	4356	2904
4	4084	3630	3439	3267	2178
5	3267	2904	2751	2614	1742
6	2723	2420	2293	2178	1452
7	2334	2074	1965	1867	1245
8	2042	1815	1719	1634	1089
9	1815	1613	1528	1452	968
10	1634	1452	1376	1307	871
11	1485	1320	1251	1188	792
12	1361	1210	1156	1089	726
13	1257	1117	1058*	1005	670
14	1167	1037	982	933	622
15	1089	968	917	871	581
16	1021	908	760	817	545
17	961	854	809	769	512
18	908	807	764	726	484
19	860	764	724	688	459
20	817	726	688	653	436
21	778	691	655	622	415
22	743	660	625	594	396
23	710	631	598	568	379
24	681	605	573	544	363
25	653	581	550	523	348
26	628	558	529	503	335
27	605	538	509	484	323
28	583	519	491	467	311
29	563	501	474	450	300
30	545	484	459	436	290

***Example:** A grower's field is 1058 long furrowed out on 38-inch centers. Therefore, every 13 rows across the field represents an acre.

QUICK CONVERSIONS

TEMPERATURE	
˚C	˚F
100	212
90	194
80	176
70	158
60	140
50	122
40	104
35	95
30	86
25	77
20	68
15	59
10	50
5	41
0	32
-5	23
-10	14
-15	5
-20	-4
-25	-13
-30	-22
-40	-40

LENGTH	
cm	inch
2 1/2	1
5	2
10	4
20	8
30	12
40	16
50	20
60	24
70	28
80	32
90	36
100	39
200	79
	feet
300	10
400	13
500	16
1,000	33

VOLUME	
liters	quarts
1	1.1
2	2.1
3	3.2
4	4.2
5	5.3
6	6.3
7	7.4
8	8.5
9	9.5

QUICK CONVERSIONS

kg./ha.	lb./A
1	0.9
2	1.8
3	2.7
4	3.6
5	4.5
10	9
20	18
30	27
40	36
50	45
60	54
70	62
80	71
90	80
100	89
200	180
300	270
400	360
500	450
600	540
700	620
800	710
900	800
1000	890
2000	1800
	ton/A
3000	1 1/4
4000	1 3/4
5000	2 1/4
6000	2 3/4
7000	3
8000	3 1/2
9000	4
10000	4 1/2
11000	5
12000	5 1/2
13000	5 3/4
14000	6 1/4
15000	6 3/4
16000	7
17000	7 1/2
18000	8
19000	8 1/2
20000	9

USEFUL MEASUREMENTS

LENGTH

1 mile = 80 chains = 8 furlongs = 1760 yards = 5280 ft. = 1.6 kilometers
1 chain = 22 yards = 4 rods, poles or perches = 100 links

WEIGHT

1 long ton = 20 cwt. = 2240 lbs.
1 lb. = 16 ozs. = 454 grams = 0.454 kilograms
1 short ton = 2000 lbs.
1 metric ton = 2204 lbs = 1000 kilograms

AREA

1 acre = 10 sq. chains = 4840 sq. yards = 43560 sq. ft. = 0.405 hectares
1 sq. mile = 640 acres = 2.59 kilometers
1 hectare = 2.471 acres

VOLUME

1 gal. = 4 quarts = 8 pints = 128 fluid ozs. = 3.785 liters
1 fluid oz. = 2 tablespoons = 4 dessertspoons = 8 teaspoons = 28.4 c.c.'s
1 liter = 1000 c.c.'s = 0.22 Imperial gallon = 1.76 pints

CAPACITIES

Cylinder-diameter^2/ x depth x 0.785 = cubic feet
Rectangle-breadth x depth x length = cubic feet
Cubic feet x 6.25 = gallons

QUICK CONVERSIONS

1 pint/acre	= 1 fluid oz./242 sq. yards
1 gal./acre	= 1 pint/605 sq. yards
1 lb./acre	= 1 oz./300 sq. yards
1 cwt./acre	= 0.37 oz./sq. yard
1 m.p.h.	= 88 ft./minute
3 m.p.h.	= 1 chain/15 sec.
1 liter/hectare	= 0.089 gal./acre
1 kilogram/hectare	= 0.892 lb./acre
1 c.c./100 liters	= 0.16 fl. oz./100 gallons
125 c.c./100 liters	=1 pint/100 gallons
1 gm./100 liters	= 0.16 oz./100 gallons

A strip 3 ft. wide x 220 chains	
A strip 4 ft. wide x 165 chains	1 acre
A strip 5 ft. wide x 132 chains	

CONVERSION FACTORS USED IN CALCULATION

Convert	To	By
gram (gm.)	kilogram (kg.)	move decimal 3 places to left
Example: 2000 gm. = 2.0 kg.		
gram (gm.)	milligram (mg.)	move decimal 3 places to right
Example: 2.0 gm. = 2000 mg.		
gram (gm.)	pound (lb.)	divide by 454
Example: 658 gm./ ÷ 454 = 1.45 lb.		
gram/pound	percent (%)	divide by 4.54
Example: 90 gm./lb. ÷ 4.54 = 19.8%		
gram/ton	percent	multiply by 11, move decimal 5 places to left
Example: 45 gm./ton x 11 = 495 = .00495%		
kilogram (kg.)	gram	move decimal 3 places to right
Example: 5.5 kg. = 5500 gm.		
milligram (mg.)	gram	move decimal 3 places to left
Example: 95 mg. = 0.095 gm.		
percent	gram/pound	multiply by 4.54
Example: 25 x 4.54 = 113.5 gm./lb.		
percent	parts/million (ppm.)	move decimal 4 places to right
Example: .025% = 250 ppm.		
percent	gram/ton	divide by 11, move decimal 5 places to right
Example: .011 ÷ 11 = .001 = 100 gm./ton		
pound	gram	multiply by 454
Example: 0.5 lb. x 454 = 227 gm.		
ppm	percent	move decimal 4 places to left
Example: 100 ppm = 0.01%		

GRAMS/GALLONS TABLE

Gallons PPM	5	10	15	20	25	50	75	100	150	200	300	400
5	0.1	0.2	0.3	0.4	0.5	1.0	1.4	1.9	2.8	3.8	5.7	7.6
10	0.2	0.4	0.6	0.8	1.0	1.9	2.8	3.8	5.7	7.6	11.0	15.0
15	0.3	0.6	0.9	1.1	1.4	2.8	4.3	5.7	8.5	11.0	17.0	23.0
20	0.4	0.8	1.1	1.5	1.9	3.8	5.7	7.6	11.0	15.0	23.0	30.0
25	0.5	0.9	1.4	1.9	2.4	4.7	7.1	9.5	14.0	19.0	28.0	38.0
50	0.9	1.9	2.8	3.8	4.7	9.5	14.0	19.0	28.0	38.0	57.0	76.0
75	1.4	2.8	4.3	5.7	7.1	14.0	21.0	28.0	43.0	57.0	85.0	114.0
100	1.9	3.8	5.7	7.6	9.5	19.0	28.0	38.0	57.0	76.0	114.0	151.0
125	2.4	4.7	7.1	9.5	12.0	24.0	36.0	47.0	71.0	95.0	142.0	189.0
150	2.8	5.7	8.5	11.0	14.0	28.0	43.0	57.0	85.0	114.0	170.0	227.0
175	3.3	6.6	9.9	13.0	17.0	33.0	50.0	66.0	99.0	133.0	199.0	265.0
200	3.8	7.6	11.0	15.0	19.0	38.0	57.0	76.0	114.0	151.0	227.0	303.0
250	4.7	9.5	14.0	19.0	24.0	47.0	71.0	95.0	142.0	189.0	284.0	379.0
300	5.7	11.0	17.0	23.0	28.0	57.0	85.0	114.0	170.0	227.0	341.0	454.0
400	7.6	15.0	23.0	30.0	38.0	76.0	114.0	151.0	227.0	303.0	454.0	606.0

STANDARD MEASUREMENTS

Measure of Length (Linear Measure)

4 inches	=	1 hand
9 inches	=	1 span
12 inches	=	1 foot
3 feet	=	1 yard
6 feet	=	1 fathom
5 1/2 yards - 16 1/2 feet	=	1 rod
40 poles	=	1 furlong
8 furlongs	=	1 mile
5,280 feet = 1,760 yards	=	320 rods = 1 mile
3 miles	=	1 league

Measure of Surface (area)

144 square inches	=	1 square foot
9 square feet	=	1 square yard
30 1/4 square yards	=	1 square rod
160 square rods	=	1 acre
43,560 square feet	=	1 acre
640 square acres	=	1 square mile
36 square miles	=	1 township

Surveyor's Measure

7.92 inches	=	1 link
25 links	=	1 rod
4 rods	=	1 chain
10 square chains	=	160 square rods=1 acre
640 acres	=	1 square mile
80 chains	=	1 mile
1 Gunter's chain	=	66 feet

Metric Length

1 inch	=	2.54 centimeters
1 foot	=	.305 meter
1 yard	=	.914 meter
1 mile	=	1.609 kilometers
1 fathom	=	6 feet
1 knot	=	6,086 feet
3 knots	=	1 league
1 centimeter	=	.394 inch
1 meter	=	3.281 feet
1 meter	=	1.094 yards

Troy Weight

24 grains	=	1 pennyweight
20 pennyweight	=	1 ounce
12 ounces	=	1 pound

Liquid Measure

2 cups	=	1 pint
4 gills	=	1 pint
16 fluid ounces	=	1 pint
2 pints	=	1 quart
4 quarts	=	1 gallon
31 1/2 gallons	=	1 barrel
2 barrels	=	1 hogshead
1 gallon	=	231 cubic inches
1 cubic foot	=	7.48 gallons
1 teaspoon	=	.17 fluid ounces (1/6 oz.)
3 teaspoons (level)	=	1 tablespoon (1/2 oz.)
2 tablespoons	=	1 fluid ounce
1 cup (liquid)	=	16 tablespoons (8 oz.)
1 teaspoon	=	5 to 6 cubic centimeters
1 tablespoon	=	15 to 16 cubic centimeters
1 fluid ounce	=	29.57 cubic centimeters

Cubic Measure (Volume)

1,728 cubic inches	=	1 cubic foot
27 cubic feet	=	1 cubic yard
2,150.42 cubic inches	=	1 standard bushel
231 cubic inches	=	1 standard gallon (liquid)
1 cubic foot	=	4/5 of a bushel
128 cubic feet	=	1 cord (wood)
7.48 gallons	=	1 cubic foot
1 bushel	=	1.25 cubic feet

Dry Measure

2 pints	=	1 quart
8 quarts	=	1 peck
4 pecks	=	1 bushel
36 bushels	=	1 chaldron

Apothecaries' Weight

20 grains	=	1 scruple
3 scruples	=	1 dram
8 drams	=	1 ounce
12 ounces	=	1 pound
27 11/32 grains	=	1 dram
16 drams	=	1 ounce
16 ounces	=	1 pound
2,000 pounds	=	1 ton (short)
2,240 pounds	=	1 ton (long)

Conversion Factors

Degree C = 5/9 (Degree F - 32).
Degree F = 9/5 (Degree C + 32).
Degrees absolute (Kelvin) = Degrees centigrade + 273.16.
Degrees absolute (Rankine) = Degrees fahrenheit + 459.69.

Multiply	By	To Obtain
Diameter circle	3.1416	Circumference circle
Diameter circle	0.8862	Side of equal square
Diameter circle squared	0.7854	Area of circle
Diameter sphere squared	3.1416	Area of sphere
Diameter sphere cubed	0.5236	Volume of sphere
U.S. Gallons	0.8327	Imperial gallons (British)
U.S. Gallons	0.1337	Cubic feet
U.S. Gallons	8.330	Pounds of water (20° C)
Cubic feet	62.427	Pounds of water (4° C)
Feet of water (4° C)	0.4335	Pounds per square inch
Inch of mercury (0° C)	0.4912	Pounds per square inch
Knots	1.1516	Miles per hour

Figuring Grain Storage Capacity

1 bu. ear corn = 70 lbs. 2.5 cu. ft. (15.5% moisture)
1 bu. shelled corn = 56 lbs. 1.25 cu. ft. (15.5% moisture)
1 cu. ft. = 1/2.50 = .4 bu. of ear corn
1 cu. ft. =1/1.25 = .8 bu. of shelled corn; Bu. x 1.25 $ft.^3$, $Ft.^3$ x .8 bu.
$Ft.^3$ = Bu. x 1.25
Bu. = $Ft.^3$ x .8
Rectangular or square cribs or bins

cu. ft. = width x height x length (W x H x L)

Round cribs, bins or silos (= 3.1416)

Volume = $R^2H = D^2H/r$

cu. ft. = x radius x radius x height = (R x R x H)

or $\frac{\text{x diameter x diameter x height} = (D \times D \times H)}{4}$

or $\frac{3.1416 \times D \times D \times H}{4} = .785 \times D \times D \times H$

Figuring Grain Storage Capacity (Continued)

Examples:

1. Crib - ear corn - 6' wide by 12' high by 40' long
   a. 6 x 12 x 40 = 2880 cu. ft. x 4 bu./cu. ft. = 1152 bu.
   b. 6 x 12 x 1 = 72 cu. ft. x .4 28.8 bu./ft. of length x 40' = 1152 bu.
2. Round crib - ear corn - 14' diameter by 14' high
   a. .785 x 14' x 14' x 14' x .8 = 1722 bushel
   b. .785 x 14' x 1 x .4 6.15 bu./ft. x 14 = 861 bushel
3. Round Bin or Silo - shell corn - 14' diameter by 14' high
   a. .785 x 14' x 14' x 14' x .8 = 1722 bushel
   b. .785 x 14' x 14' x 14' x 1 x .8 - 123 bu./ft. x 14' = 1722 bushel

Metric Equivalents

Metric Weight

1 grain	=	.065 gram
1 apothecaries' scruple	=	1.296 grams
1 avoirdupois ounce	=	28.350 grams
1 troy ounce	=	31.103 grams
1 avoirdupois pound	=	.454 kilogram
1 troy pound	=	.373 kilogram
1 gram	=	15.432 grains
1 gram	=	.772 apothecaries' scruple
1 gram	=	.035 avoirdupois ounce
1 gram	=	.032 troy ounce
1 kilogram	=	2.205 avoirdupois pounds
1 kilogram	=	2.679 troy pounds

Capacity

1 U.S. fluid ounce	=	29.573 ml
1 U.S. fluid quart	=	.946 liter
1 U.S. liquid ounce	=	29,573 milliliters
1 U.S. liquid quart	=	.964 liter
1 U.S. dry quart	=	1.101 liters
1 U.S. gallon	=	3.785 liters
1 U.S. bushel	=	.3524 hectoliters
1 cubic inch	=	16.4 cubic centimeters
1 liter	=	1,000 milliliters or 1,000 cubic centimeters
1 cubic foot water	=	7.48 gallons or 62 1/2 pounds
231 cubic inches	=	1 gallon
1 millimeter	=	.034 U.S. fluid ounce
1 liter	=	1.057 U.S. liquid quarts
1 liter	=	.908 U.S. dry quart
1 liter	=	.264 U.S. gallon
1 hectoliter	=	2.838 U.S. bushels
1 cubic centimeter	=	.061 cubic inch

Miscellaneous Equivalents

9 in. equals 1 span

6 ft. equals 1 fathom

6,080 ft. equals 1 nautical mile

1 board ft. equals 144 cu. in.

1 cylindrical ft. contains 5 7/8 gals.

1 cu. ft. equals .8 bushel

12 dozen (doz.) equals 1 gross (gr.)

1 gal. water weighs about 8 1/3 lbs.

1 gal. milk weighs about 8.6 lbs.

1 gal. cream weighs about 8.4 lbs.

46 1/2 qts. of milk weighs 100 lbs.

1 cu. ft. water weighs 62 1/2 lbs., contains 7 1/2 gals.

1 gal. kerosene weighs about 6 1/2 lbs.

1 bbl. cement contains 3.8 cu. ft.

1 bbl. oil contains 42 gals.

1 standard bale cotton weighs 480 lbs.

1 keg of nails weighs 100 lbs.

4 in. equals 1 hand in measuring horses

ADDRESS OF BASIC MANUFACTURERS

1. Abbott Laboratories
   Abbott Park-1400 Sheridan Road
   N. Chicago, IL 60064

2. ABH Chemie Export/Import
   Volkseigener Aussenhandelsbetrieb
   der DDR
   Storkower Strasse 133
   DDR - 1055 Berlin
   East Germany

3. Agrimont Spa
   Montedison Group
   Piazza Della Republica 6
   20 124 Milano, Italy

4. Agrisence
   83 E. Shaw # 250
   Fresno, CA 93710

5. Agrolinz
   Agrarchemihalein GmbH
   St. Peter Strasse 25
   POB 21
   A-4021 Linz, Austria

5. Agtrol Chemical Products
   7324 Southwest Freeway Ste 1800
   Houston, TX 77074

6. Agway
   P.O. Box 4933
   Syracuse, NY 13221

7. Akzo Zout Chemie Nederland by
   Postbus 4080
   1009 AB Amsterdam
   The Netherlands

8. American Cyanamid Co.
   One Cyanamid Plaza
   Wayne, NJ 07470

9. Applied Biochemists
   Box 255
   Mequion, WI 53092

10. Asahi Chemical Industry Co.
    Hibiya-Mitsui Bldg.
    1-2, Yurako Cho 1-Chome
    Chiyoda-ku, Tokyo 100
    Japan

11. Atlas Interlades, Ltd.
    Fraser Rd, Erith
    Kent DA8 1PN
    England

12. AVITROL Corp.
    320 S. Boston
    Ste. 528
    Tulsa, OK 74103

13. BASF
    Postfach 220
    7603 Limburgerhof
    West Germany (FR)

14. BASF-Wyandotte Corp.
    Ag Chemical Div.
    100 Cherry Hill Road
    Parsippany, NJ 07054

15. Bayer AG
    5090 Leverkusen
    Bayerwerk,
    West Germany (FR)

16. Bayvet Division-Miles Labs
    P.O. Box 390
    Shawnee, KS 66201

17. Bell Labs
    3699 Kinsman Boulevard
    Madison, WI 53704

18. Buckman Labs
1256 N. McLean Blvd.
Memphis, TN 38119

19. C. P. Chemicals, Inc.
P.O. Box 158
Sewaren, NJ 07077

20. Cheminova
P.O. Box 9
DK-7610 Lemvig,
Denmark

21. Chemolimpex
Plant Protection Dept.
P.O. Box 121
H-1805 Budapest,
Hungary

22. Chempar
P.O. Box 09186
Milwaukee, WI 53209

23. Chemserv Industrie Service GmbH
CE-DB Bau 8
Postfach 296
A-4021 Linz, Austria

24. Chugai Pharmaceutical Co. Ltd.
1-9 Kyobashi 2-Chome, Chuo-ku
Tokyo 104 Japan

25. CIBA-Geigy AG
CH-4002 Basel 7
Switzerland

26. CIBA-Geigy Corporation
Agricultural Chemicals
P.O. Box 18300
Greensboro, NC 27419

27. W.A. Cleary Company
P.O. Box 10
Somerset, NJ 08873

28. Cooper McDougall &
Robertson, Ltd.
Berkhamsted, Herts.
England

29. d-Con Company
90 Park Avenue
New York, NY 10016

30. Dainippon Ink & Chemicals
DIC Building
7-20 Nihonbashi 3 Chome
Chuo-ku
Tokyo 103 Japan

31. Degesch America, Inc.
P.O. Box 116
Weyers Cave, VA 24486

32. Degesch GmbH
P.O. Box 610207
D-6000 Frankfurt 61
West Germany (FR)

33. Dow Chemical Co.
P.O. Box 1706
Midland, MI 48640

34. Dow Chemical Ltd.
Latchmore Court, Brand St.
Hitchin, Herts SG5 1H2
England

35. Dr. R. Maag Ltd.
Chemical Works
CH-8157
Dielsdorf, Switzerland

36. Drexel Chemical Co.
P.O. Box 9306
Memphis TN 38109

37. Dunhill Chemical Co.
3026 Muscatel Ave.
Rosemead, CA 91770

38. Duphar BV
Crop Protection Div.
P.O. Box 900
1380 DA Weesp
Holland

39. DuPont Company
Agricultural Chemical Dept.
Barley Mill Plaza
Wilmington, DE 19898

40. Eastman Kodak Co.
Life Sciences Div.
343 State Street
Rochester, NY 14650

41. Ecogen
2005 Cabot Blvd., West
Langhorn, PA 19047-1810

42. EM Laboratories
5 Skyline Drive
Hawthorne, NY 10532

43. Elanco Products Company
Division of Eli Lilly Co.
Lilly Corporate Center
Indianapolis, IN 46285

44. Evans Bio Control, Inc.
895 Interlocken Pkwy. Unit A
Broomfield, Co 80020

45. FMC Corporation
Ag Chemical Div.
2000 Market Street
Philadelphia, PA 19103

46. F. W. Berk & Co., Ltd.
8 Baker Street
London W1, England

47. Fairfield American
201 Route 17 North
Rutherford, NJ 07070

48. Farmoplant Spa
Piazza Della Republic 6
20124 Milano, Italy

49. Fermenta Plant Protection
5966 Heisley Rd.
Mentor, OH 44061-8000

50. Fermone Chemicals, Inc.
1700 N. 7th Ave., Ste 100
Phoenix, AZ 85007

51. Fujisawa Pharmaceutical
4-3 Doshomachi 4-Chome
Higashiku Osaka 541
Japan

52. Gilmore, Inc.
1755 N. Kirby Parkway
Ste 300
Memphis, TN 38119

53. Great Lakes Chemical Corp.
P.O. Box 2200
W. Lafayette, IN 47906

54. Griffin Corporation
P.O. Box 1847
Valdosta, GA 31601

55. Gustafson, Inc.
1400 Preston Rd., Ste. 400
Plano, TX 75266

56. Guth Corporation
551 Grandville
Hillside, IL 60162

57. Hercon
200 B Corporate Ct.
Middlesex Business Center
Plainfield, NJ 07080

58. Hodogaya Chemical
1-4-2 Toranomon-1-Chome
Minato-ku
Tokyo 105 Japan

59. Hoechst-Agrochem. Div.
Postfach 80 03 20
6230 Frankfurt (m) 80
West Germany (FR)

60. Hoechst-Roussel Agri.
Vet. Co.
Route 202-206 North
Somerville, NJ 08876

61. Hokko Chemical Industries
Mitsui Building 2
4-2 Nihonbashi Hongoku-cho
Tokyo 103 Japan

62. Hopkins Ag Chemical Co.
P.O. Box 7532
Madison, WI 53707

63. ICI Americas, Inc.
Ag Chemical Division
Wilmington, DE 19897

64. ICI Ltd.
Plant Protection Div.
Fernhurst, Hasslemere
Surrey, England GU26 3JE

65. Ihara Chemical Co.
1-4-26 Ikenohata 1-Chome
Taitoku
Tokyo 110 Japan

66. Ishihara Sangyo Kaisha Ltd.
10-30 Fujimi 2-Chome
Chiyoda-ku, Tokyo
Japan

67. Janssen Pharmaceutical
P.O. Box 344
Washington Crossing, NJ 08560

68. Janssen Pharmaceutica N. V.
Agricultural Div.
B-2340 Beerse, Belgium

69. Kaken Chemical Co., Ltd.
No. 28-8, 2 Chome
Honkomagome, Bunkyo-ku
Tokyo 113 Japan

70. Kalo Labs
4550 W. 109th St.
Suite 222
Overland Park, KS 66211

71. Kanesho Co., Ltd.
Rm. 333, Marunouchi Bldg.
Maunouchi, Chiyoda-ku
Tokyo, Japan

72. Keno Gard AB
P.O. Box 11033
S-10061 Stockholm, Sweden

73. Kumiai Chemical Industries
4-26 Ikenohata 1-Chome
Tokyo 110 Japan

74. Kureha Chemical Ind. Co.
1-9-11 Nihonbashi,
Horidome-cho, Chuo-ku,
Tokyo 103 Japan

75. Lipha
115 Avenue Lacassagone
69003 Lyon, France

76. Maag Agrochemicals
P.O. Box 2048
Vero Beach, FL 32960

77. Maktheshim - Agan
P.O. Box 60
Beer-Sheva, Israel

78. McLaughlin Gromley King
8810 - 10th Ave., North
Minneapolis, MN 55427

79. Meiji Seika Company
4-16 Kyobashi 2-Chome
Chuo-ku
Tokyo 104 Japan

80. Merck & Company
Sharpe & Dohme Research Lab.
Hillsborough Rd.
Three Bridges, NJ 08887

81. E. Merck A. G.
61 D Armstadt
Franfurter Strasse 250
West Germany (FR)

82. Miller Chemical & Fert. Corp.
Box 333
Hanover, PA 17331

83. Minerals Res. & Devel. Corp.
4 Woodlawn Green
Suite 232
Charlotte, NC 28210

84. Mitsubishi Chemical Ind. Ltd.
Agric. Chemical Div.
Mitsubishi Shozi Bldg.
6-3, Marunouchi 2-Chome
Chiyoda-ku
Tokyo, Japan

85. Mitsubishi Petrochemical Co. Ltd.
2-5-2 Marunouchi, 2-Chome
Chiyoda-ku
Tokyo 100 Japan

86. Mitsui Agricultural Chemicals
1-2-1 Ohtemachi, Chiyoda-ku,
Tokyo, Japan

87. Mitsui Toatsu Chemicals
P.O. Box 83
Kasumigasehi Building
Tokyo 100 Japan

88. Mobay Chemical Co.
P.O. Box 4913
Kansas City, MO 64120

89. Monsanto Chemical Company
800 N. Lindburgh Blvd.
St. Louis, MO 63167

90. Motomco Ltd.
P.O. Box 6072
Clearwater, FL 33518

91. MSD Ag Vet.
P.O. Box 2000
Rahway, NJ 07065

92. Mycogen Corp.
5451 Oberlin Dr.
San Deigo, CA 92121

93. Nalco Chemical Company
2901 Butterfield Rd.
Oak Brook, IL 60521

94. Nihon Nohyaku Company, Ltd.
2-5 Nihonbashi 1-Chome
Chuo-ku
Tokyo 103 Japan

95. Nihon Takushu Noyaku Seizo KK
Honcho Bldg., 2-4 Nihonbashi
Honcho, Chuo-ku
Tokyo 103 Japan

96. Nippon Kayaku Co.- Ag Div.
Tokyo Kaijo Bldg.
2-1, Marunouchi 1-Chome
Chiyoda-ku
Tokyo 100 Japan

97. Nippon Soda Co., Ltd.
New-Ohtemachi Bldg.
2-1, 2-Chome Ohtemachi
Chiyoda-ku
Tokyo 100 Japan

98. Nissan Chemical Ind., Ltd.
Kowa-Hitotsubashi Bldg.
7-1, 3-Chome, Kanda
Nishiki-cho Chiyoda-ku
Tokyo 101 Japan

99. Nor-Am Ag. Products, Inc.
3509 Silverside Rd.
Wilmington, DE 19803

100. Nordox A/S
Ostenjovn 13
Oslo 6 Norway

101. Novo Biokontrol
33 Turner Road
Danbury CT 06810-5101

102. Otsuka Chemical Co.
10 Bungo-Machi
Higashi-ku
Osaka 540 Japan

103. Pennwalt Corporation
Ag Chemical Div.
3 Parkway
Philadelphia, PA 19102

104. Pepro
B. P. 139
69 Lyon R. P.
France

105. Pestcon Systems
P.O. Box 469
Alhambra, CA 91802

106. Pfizer Inc. - Chemical Div.
235 E. 42nd St.
New York, NY 10017

107. Phelps Dodge Refining Corp.
300 Park Avenue
New York, NY 10022

108. Phillips Petroleum
Bartlesville, OK 74004

109. Prentiss Drug &
Chemical Co., Inc.
21 Vernon St. CB 2000
Floral Park, NY 11001

110. Proctor & Gamble
Cincinnati, OH 45202

111. Ralston Purina Company
Checkerboard Square
St. Louis, MO 63188

112. Rentokil Laboratories
Felcourt, East Grinstead
Sussex, England

113. Rhone Poulenc
37-39 Manor Rd.
Romford, Essex
2MI 2TL England

114. Rhone Poulene
P.O. Box 12014
2 TW Alexander Dr.
Research Triangle Park, NC 27709

115. Rhone Poulenc Agrochemie
14-20 rue Pierre Baizet
69009 Lyon, France

116. Rohm & Haas Company
Independence Mall West
Philadelphia, PA 19105

117. Roussel Bio Corp.
400 Sylvan Ave.
Englewood Cliffs, NJ 07632

118. Roussel UCLAF
163, Ave. Gambetta
75020 Paris, France

119. Sandoz Crop Protection Corp.
1300 E. Touhy Ave.
Des Plaines, IL 60018

120. Sandoz, Ltd.
Agrochemical Dept.
Basel, Switzerland CH-4002

121. Sankyo Co., Ltd.
No. 7-12, Ginza, 2-Chome
Chuo-ku
Tokyo 104 Japan

122. Sapporo Breweries, Ltd.
10-1 Ginza 7-Chome
Chuo-ku,
Tokyo 104 Japan

123. SARIAF
20124 Milano
Italy

124. Schering AG
Hauxton Cambridge
CB2 5HU England

125. Schering AG
Postfach 650311
Berlin 65,
West Germany (FR)

126. SDS Biotech KK
No.2 Higashi Shinbashi Bldg.
12-7 Higashi Shinbashi 2- Chome
Minato-ku
Tokyo 105 Japan

127. Shell Agrar
P.O. Box 202
6507 Ingelheim
West Germany (FR)

128. Shell International Ltd.
Agrochemical Div.
Shell Centre
London SE1 7PG
England

129. Shionogi and Co., Ltd.
12- Doshomachi 3-Chome
Osaka, 541 Japan

130. Sierra Chemical Co.
1001 Yosemite Dr.
Milpitas, CA 95035

131. Sumitomo Chemical Americas
1330 Dillon Heights Ave.
Baltimore, MD 21258

132. Sumitomo Chemical Co., Ltd.
15- 5 Chome Kitahama
Higashi-ku,
Osaka 541 Japan

133. Takeda Chemical Industries
12-10 Nihonbashi 2-Chome
Chuo-ku
Tokyo 113 Japan

134. Tennessee Chemical Co.
3475 Lenox Rd., NE
Suite 670
Atlanta, GA 30326

135. 3-M Company
Agricultural Chemical
Products Bldg. 223
3-M Center
St. Paul, MN 55144

136. Tosoh Corporation
1-7-7 Akasaka
Minato-ku,
Tokyo 107 Japan

137. UBE Ind. Ltd.
Ark Miri Bldg.
12-32 Akasaka 1-Chome
Minato-ku
Tokyo 107 Japan

138. Uniroyal Chemical Co.
Crop Protection Division
Middlebury, CT 06749

139. Unocal Chemical Corp.
P.O. Box 60455
Los Angeles, CA 90060

140. Valent Corp.
1333 No. California St.
Walnut Creek, CA 94598

141. Vineland Chemical Company
West Wheat Road
Vineland, NJ 08360

142. Wacker-Chemie GmbH
Prinzregentenstr. 22,
800 Munchen 22,
West Germany (FR)

143. Zoecon Corporation
975 California Avenue
Palo Alto, CA 94304

NOTES

NOTES

NOTES

NOTES